Systematic Data Analysis and Reporting

An Introduction to the Craft of Making Your Analytic Work "Bulletproof"

By

Daniel R. Bretheim

First published by AuthorHouse 08/15/05

ISBN: 1-4184-7210-7 (e-book)
ISBN: 1-4184-2733-0 (Paperback)

Library of Congress Control Number: 2004092121

This book is printed on acid free paper.

Printed in the United States of America
Bloomington, IN

SAS® and all other SAS Institute Inc. product or service names are registered trademarks or trademarks of SAS Institute Inc. in the USA and other countries. ® indicates USA registration.

Material from Combining the Data, Chapter 5, taken from the e-Learning course, SAS® Certification Training: Core Concepts, Version 8, copyright ©2000, SAS Institute Inc., Cary, NC, USA. All Rights Reserved. Reproduced with permission of SAS Institute Inc., Cary, NC

IBM® and DB2® are trademarks of International Business Machines Corporation in the United States, other countries, or both.

Oracle® and all other Oracle Corporation product service names are registered trademarks or trademarks of Oracle Corporation in the USA and other countries.

LEGO® and all other LEGO A/S Corporation product names are registered trademarks or trademarks of the LEGO Group.

Microsoft® is a trademark of Microsoft Corporation in the United States, other countries, or both.

Lotus Notes® is a trademark of Lotus Development Corporation.

Acknowledgements

A number of colleagues over the years have helped shape the concepts that appear in this book (whether they realize it or not!). In doing so they have also contributed, each in their own way, to the satisfaction I enjoy from working with data.

Lance Heineccius opened the door to my first job as a programmer/analyst. His patience and my persistence dovetailed well. Ron Liebman is probably the smartest analyst I've ever had the pleasure to work with. I've always brought my toughest data challenges to him. Ryan Carr is, in my opinion, the "Michael Jordan" of SAS programming. Working with him forced me to be a better programmer. Dr. Charlie Barr has been a wonderful resource for brainstorming about how to manage the analytical process in the workplace. I envy his deep and broad skill set.

Finally, I'd like to thank Solucient for allowing me to use the health care claims data represented in Case Study II.

Table of Contents

Preface

I wish that there had been a course in college or graduate school called "Data Analysis 101". I have taken a number of statistics and computer programming courses that have taught me specific tools. But there was no course available to help prepare me for organizing my work as a programmer/analyst. I understood analytical issues and I knew how to extract information from data, but the missing component was a systematic work process that would structure the analytical activity in a way that would increase my efficiency, document the results, and ensure that I could replicate the results, should that be required. Therefore, I found it necessary to develop an approach and set of standards that would organize my work so that the following questions could be answered:

- Who - needs this information?
- What - questions do they want answered?
- When - do they need the answers?
- Where - are the data?
- Why - is this important?
- How - did I get the answer?

It is the "How" question above that will be the focus of most of this book. Chapters 1 through 9 provide an overview of my vision for a systematic approach to analytic work. The Appendix, which contains three comprehensive (although relatively simple) case studies is the real "heart and soul" of the book. The case studies apply the process using data representative of three different functional areas. It is my belief that case studies bring concepts alive and thereby stimulate the imagination of the reader to take those concepts and implement/tailor them in the unique context of their own work environment.

Throughout this book you will see many references to the SAS® System. Please note that this is not a book about SAS programming. There are many other excellent resources devoted to that topic. The SAS System is used to demonstrate concepts and tools that are discussed in this book. Other software tools could have been used. I chose the SAS System because I believe it is an excellent tool. However, you can apply these concepts using any tool set.

Not all analysts will choose to use a systematic approach to data analysis. For those that do, the decision is the easy part. Following through on that decision and developing the discipline to apply the process is actually very challenging, particularly when under the pressure of tight project deadlines. With that reality in mind, I have attempted to develop a process approach to analytic work that is simple yet comprehensive in scope, flexible in how it can be implemented, and practical, i.e., the benefit of using a systematic approach is greater than the costs of doing so. If you choose to adopt a systematic approach to your work, please remember that it is truly is a craft, and therefore takes time and practice to master. The first time you attempt to apply these concepts you will be frustrated. Persevering through that initial experience becomes a matter of faith that in the end, the effort will have been worthwhile. Over the past twenty years I have become increasingly convinced that a systematic approach to analytic work has a significant payback, to me professionally as well as for my client.

1. The Process

Introduction

This chapter introduces a process model for analytic work. Each component of the model is briefly described, including the relationship and interaction between components. In the chapters that follow, each component is described in further detail.

What is Data Analysis?

In my opinion, most if not all data analysis is undertaken because someone has a question and is looking for an answer. Getting from question to answer usually involves multiple sequential steps, often with repetition of steps until the answer is produced. This means that data analysis is a process with inputs, throughputs, and outputs that can iterate. As with any process, it can be designed and executed to produce results efficiently and accurately. Because it is a process executed and managed by humans, it is carried out using a variety of learned skills and tools. Our ability as analysts to conduct this work efficiently and effectively indicates the level of our proficiency, or craft. Therefore, I distinguish the process of data analysis from the set of tests (e.g., hypothesis testing), skills (e.g., programming), and tools (e.g., application software) that are used by the analytic process. This book will not teach you how to be a statistician or programmer, or how to design research studies. However, it will impart an understanding of how data analysis can be viewed as a process that can be optimized to efficiently produce accurate and replicable results.

The Value of Following a Defined Process

Some analysts bristle at the thought of using a defined process to organize and conduct their work. It's often perceived as too constraining or cumbersome, amounting to nothing more than administrative overhead. "How can you expect me to get work done under these aggressive deadlines if I have to follow a process?" is a common reaction. "I can't work fast if you require me to follow a process" is another complaint.

My response to these objections is to counter with examples of similar situations where the use of process can be the difference between success and failure. Consider the analogy of a commercial airline pilot. Written procedures are used for each phase of flight, e.g., taxi, takeoff, climb, cruise, descent, approach, and landing, to ensure that the sequence of tasks and dependencies is executed at the appropriate time and under the proper conditions. For example, at a specific point in the landing checklist the pilot lowers the landing gear. It needs to happen at the right time, e.g., (before touchdown) but not too early (e.g., if you are going too fast.) The intent is to allow the pilot to focus on flying the plane and not be distracted by trying to remember what step in the sequence comes next. This minimizes the potential for human error and creates a more efficient cockpit environment. This is particularly true in unplanned flight situations, such as an emergency, where energy must be devoted to rapid problem diagnosis and problem solving. Emergency checklists are used to ensure that important steps will not be overlooked or forgotten when things get exciting.

Therefore, whenever the sequence and timing of events is important to ensure that critical steps are not overlooked, a process can add value, even when under the pressure of an "emergency" situation.

A Process Model

Figure 1.1 is a representation of the process I use for my own analytic work. It suggests a certain sequence of core activities, all of which are enveloped in an ongoing need to stay organized. I work this way because it helps to ensure my ability to replicate the results that I've obtained. While not covered in the scope of this book, all of this activity is influenced by basic project management skills related to communication, accountability, initiative, and service to your customer. The core of the model is five sequential activities that can often iterate many times in the course of a single project. Interjected in and around these sequential activities are the ongoing activities that occur throughout the project -- getting/staying organized and documenting the results.

Figure 1.1

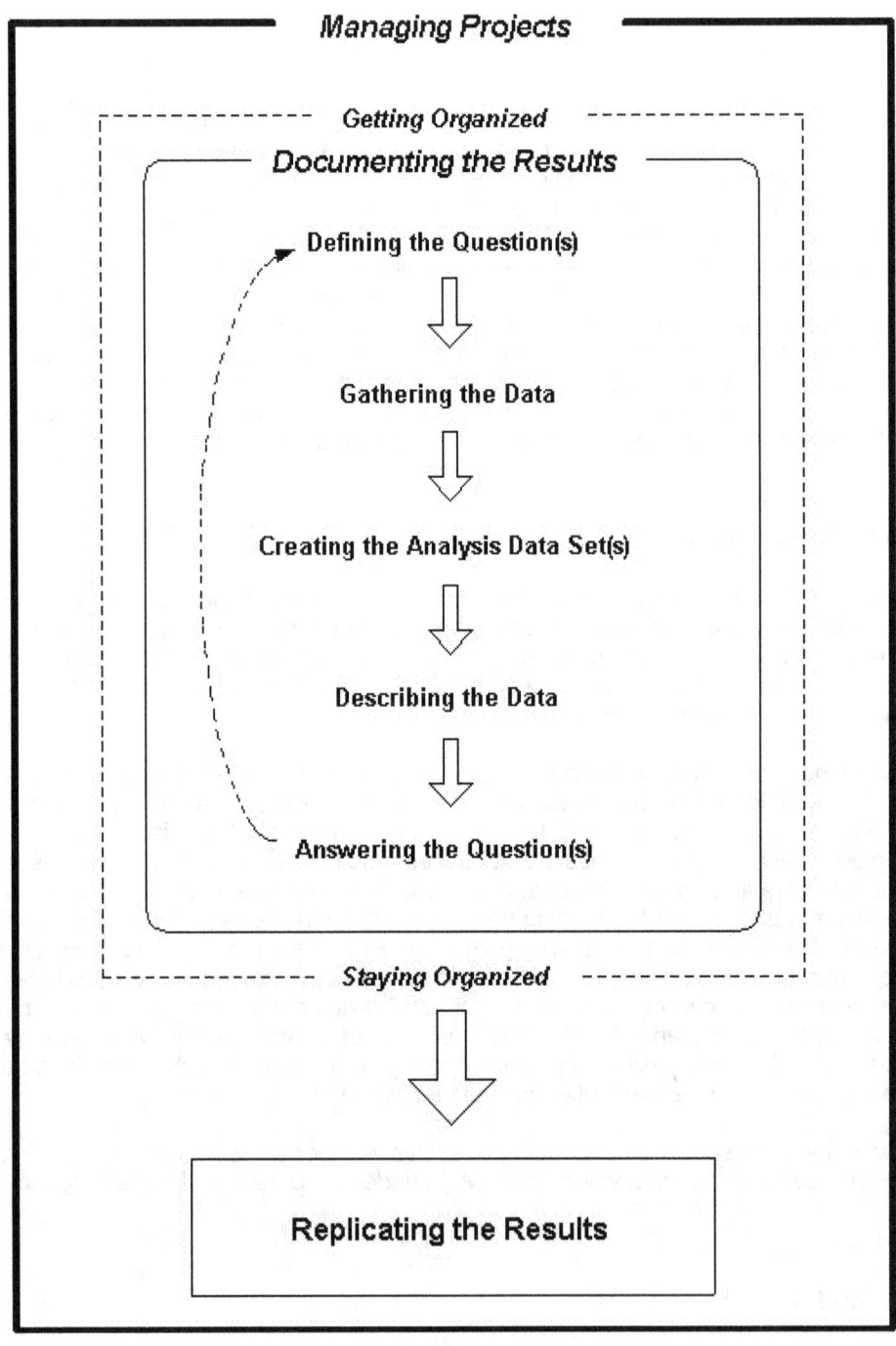

Defining the Questions

The only reason we have an analysis project is because somebody has one or more questions to be answered. Without those questions there would be no need for the project. Developing an early and thorough understanding of the questions we will answer is one of the most critical steps in the process. Inadequacies and shortcomings in this step often result in rework and lost credibility. How many times have we found ourselves in receipt of report requirements via email, voice mail, or "back of the envelope" notes? Often we're all too eager to start our project on the basis of cryptic one-way communications from our customer. Taking the time, up front, to define the questions we will answer is a worthwhile investment. It provides us the opportunity to identify and document important considerations such as assumptions, and even the more mundane details such as report titles. But perhaps the best reason for using a systematic approach for defining the questions is the opportunity to shape client expectations by establishing a realistic understanding of the level of effort and complexity that need to be factored into the delivery time frame.

Getting Organized

Before forging ahead with the next step in the process, take some time to get organized. One of the hardest things to do at the beginning of a project is to take the time to gather your thoughts, think about how the project will unfold, and set yourself up for success by organizing what you're going to do. In the heat of battle, we're anxious to move forward toward the objective in the spirit of "ready-fire-aim". It takes considerable discipline to get organized before you start to write code, i.e., "ready-aim-fire". The adage "You can pay me now or you can pay me later, but you're going to pay me" applies at this stage of an analysis project.

Gathering the Data

Truly understanding the questions to be answered positions us to obtain and process the appropriate data. Achieving the gold standard of "Getting it right the first time" is absolutely dependent on having the proper data. This step focuses on the logistics of getting data quickly and efficiently. Identifying who to contact, the options that may exist regarding format and delivery medium, how to specify your request, and jointly agreeing on when the data will be available, can significantly lessen the amount of time devoted to this stage of the project.

Creating the Analysis Dataset(s)

This step typically involves a great deal of data reading, manipulation, and output. It is the Extract-Transform-Load part of the project. As such it presents plenty of opportunities for error, particularly in relation to any assumptions we still have regarding the data. The analysis dataset(s) created during this step form the foundation for everything that will be delivered from the project.

Describing the Data

Before jumping into the analysis dataset(s) to produce the final reports, we use this step to scrutinize the data in a variety of ways. We look for extreme and unusual values, test whatever assumptions we still have, and otherwise examine the data to assess their completeness, accuracy, and validity. Only after convincing yourself that these three attributes are accounted for are you in a position to proceed with final reporting.

Answering the Questions

Even though we have a solid foundation for our analysis by this stage of the project, we may still need to build upon that foundation through additional data manipulation. This can take the form of creating additional variables, often using complex logic and data transformations. We might also find ourselves summarizing the data or reformatting it for reporting purposes. There are many potential pitfalls at this stage of the project, not to mention that we are nearing the end of the process and are probably up against a tight deadline and more than likely fatigued by this time.

3

Another possible outcome at this stage of the process is that as initial answers begin to surface and are presented to your client, you may find that the questions themselves actually change. This is why the process model (Figure 1.1) contains an arrow back to previous steps. This should not come as a surprise to an experienced analyst since analysis, almost by definition, is iterative.

Staying Organized

This activity is an extension of "Getting Organized". Given the iterative nature of data analysis, staying organized can be a challenge. "Staying organized" is not a discrete step in the process and should not be confused with the tasks associated with producing final project documentation. The need to maintain a certain level of control over your activity is faced at each stage of the process; it's an ongoing activity that is integral to building and maintaining the audit trail that links all elements of a project.

Documenting the Results

When you have successfully navigated your way through a sea of data and answered the questions posed by your client, it's time to package up the final project documentation. If you've been equally successful in staying organized throughout the course of the project, then the level of effort necessary to prepare final project documentation will be lessened. It can also be said that strong project management skills applied during the project will also contribute positively to the quality and completeness of project documentation. In the end, project documentation is the key to your ability to substantiate, defend, and replicate your results.

Comparison to Other "Models"

In thinking about this book and reviewing other works, I've come across several different perspectives about how to approach the work of data analysis. Two examples are outlined and contrasted below.

The first model is described in the Calvert and Ma book "Concepts and Case Studies in Data Management", where their focus is oriented toward pharmaceutical research and development. Their Research Data Management (RDM) model includes eight areas of activity, which I have mapped to my process model for Systematic Data Analysis and Reporting (SDAR).

SDAR	Research Data Management
Defining the Questions	
Getting Organized	7. Documentation procedures and standards
Gathering the Data	1. Data acquisition
Creating the Analysis Dataset(s)	2. Data verification and validation
Describing the Data	3. Data manipulation and analysis
Answering the Questions	4. Result reporting
Staying Organized	6. Monitoring procedures
Documenting the Results	5. Back up and archiving
Managing Projects	8. Teamwork (training, communication, support)

The second model is described in the Forsberg et al book "Visualizing Project Management", where the focus is oriented toward application development. Their "Situational Project Management Elements" model includes ten components, which are also mapped to my SDAR model as follows.

SDAR	Project Management Elements
Defining the Questions	1. Requirements
Getting Organized	2. Organizing
Gathering the Data	4. Planning
Creating the Analysis Dataset(s)	5. Opportunity and Risk Management
Describing the Data	7. Visibility
Answering the Questions	8. Status
Staying Organized	6. Project Control

Documenting the Results	
Managing Projects	3. Project Team
	9. Corrective Action
	10. Leadership

While there are some differences in approach and emphasis between these models, they share a significant amount of common ground. Therefore, rather than debate the pros and cons of different models, it's far more important that you select a model, any model, and then use it with conviction.

Investing in Documentation

Here is a statement that we can all agree with: "Creating documentation takes time and effort." There is no doubt that there is a cost associated with creating and maintaining good project documentation. I believe that this cost can legitimately be seen as an investment. If so, then we would expect a positive payback. I have found the following benefits associated with my personal investment in project documentation:

- Optimum restart points in the data flow are easily identified.
- Less wasteful rework.
- Ease of maintenance or modification.
- Transference among staff.
- The ability to replicate your results.
- In my opinion, one of the most valuable by-products of good project documentation is the confidence that comes from knowing you can explain how your results were produced.

I also feel strongly that the most efficient method for creating project documentation is to build it as you work, i.e., make it an ongoing part of your work process rather than an end of project chore. This takes discipline and awareness. Waiting until the end of a project before starting the documentation effort increases the likelihood that shortcuts will be taken, key details overlooked, or documentation will be skipped entirely.

Guiding Principles

Over the years I have concluded that there are certain basic truths that govern the world of data analysis. In response to these perceptions I have developed a list of "guiding principles" that I try to follow as consistently as possible. I have found that my belief in the validity of these statements provides a source of motivation for adhering to the SDAR process. These statements constitute the underlying creed from which SDAR has evolved. Whether you agree or disagree with these statements, being aware of them provides a useful context for understanding the rationale for SDAR.

If It Isn't Written Down It Was Never Said Or Done

While I certainly appreciate the flexibility and convenience of a "handshake agreement", I am much more comfortable in the sincerity and commitment of my counterpart when he or she will actually sign a statement of work. For that reason, I am a strong believer in the value of written agreements, whether they are related to defining a project scope (as in chapter 2), requesting data (chapter 4), or managing change (chapter 8). If your counterpart is reluctant to substantiate their commitment in the form of a signature, then you have good reason to question their buy-in to the project.

Identify and Document Assumptions

Assumptions are like land mines. They lurk under the surface of every project and are very nasty if you step on one. Your job is to ferret them out and document them. Even though some assumptions may go undetected, an effort should be made to identify as many as possible early in the project.

Get Organized

Spending a few minutes (it really is minutes, not hours, or days) to get organized before launching into a project will save you time in the long run. Deciding on naming conventions for programs, variables, and files and setting up folders to organize your data, code, and output is the starting point for building your audit trail. Take the time to do it before you start writing code.

Look For Problems Early

I think it's safe to say that much of the data we work with are inherently riddled with problems. If you don't know this to be true, please take my word for it. These problems can take the form of missing values, missing time periods, truncated values, out of range values, and invalid values, just to make a few. Your challenge is to develop the practice of searching for problems early in the project. It's part of the assessment of completeness, accuracy, and validity discussed in chapters 5 and 6.

Seek Out Master Data and Use It

Not all analysts are familiar with the concept of "master data". By "master data" I am referring to centrally maintained data sets that contain values that can and should be applied universally across the organization. Examples include data elements such as product codes, customer names, vendor addresses, foreign exchange coefficients, etc. If these master data are centrally maintained within your organization and are available to you, take full advantage of it. It will save you time and eliminate one potential source of variation between your results and someone else's. The consistent use of master data is a key factor in establishing "one version of the truth".

Stay Organized

Getting organized is one thing, staying organized is another. Your ability to stay organized during the course of an intense analysis will greatly streamline the effort required to maintain an audit trail (as discussed in chapter 8). It's analogous to stopping to tie your shoe during a race. Even though it takes some time, it increases the probability that you will successfully complete the race without tripping.

Archive Data As If Your Career Depended On It

Some of you might accuse me of being overly dramatic with this statement. However, consider the following scenario. If you had been the lead analyst on an important project and were asked to quickly replicate the study or reopen a research question, only to find out that you were unable to deliver due to inadequate documentation, how would that impact your next performance review? Comprehensive project documentation should include a record of the research questions, data analysis specifications, code, test output, final reports, conclusions, input files, and output files. The availability of these items "off the shelf" equips you to quickly and efficiently deliver results.

Documentation Should Pass the "Hit By a Bus" Test

Here's another classic scenario. If a bus hit you, could another analyst step in and finish the project or replicate your results? Your boss certainly hopes so, just as you would if you ever become the other analyst assigned to pick up the pieces.

Expect To Be Questioned -- Possibly Doubted

Information is power. If someone doesn't like your results, they might question your process, in an effort to cast doubt on your conclusions. Easy targets include any data manipulation, variable selection, or calculations that you have performed. The reason for creating "bulletproof" work is to be able to stand up to this kind of scrutiny. You must be able to demonstrate that your results have accurately addressed the questions you have been tasked to answer.

Analysis Is Iterative

Don't expect to be done the first time you present results. Expect to go back and rework the data. Have an audit trail that makes rework quick and efficient.

Confidence Comes From Knowing You Can Replicate Your Results

As I've said before, being able to clearly and confidently state how you obtained a result adds enormous credibility to your presentation.

All of This Is Easier Said Than Done

Finally I have to acknowledge that it takes a great deal of discipline and commitment to follow these principles. At some point you have to become convinced that the effort to use a process has an adequate payback.

Guiding Principles

If It Isn't Written Down It Was Never Said Or Done
Identify and Document Assumptions
Get Organized
Look For Problems Early
Seek Out Master Data and Use It
Stay Organized
Archive Data As If Your Career Depended On It
Documentation Should Pass the "Hit By a Bus" Test
Expect To Be Questioned -- Possibly Doubted
Analysis Is Iterative
Confidence Comes From Knowing You Can Replicate Your Results

Summary - Key Points

- There is value in following a defined process, but ultimately you have to convince yourself of that.
- Select a model, any model, and then use it with conviction.
- Documentation is an organizational investment.
- All of this is easier said than done.

2. Defining the Questions

Introduction

The only reason we have an analysis project is because somebody has one or more questions to be answered. Without those questions there would be no need for the project. Developing an early and thorough understanding of the questions we will answer is one of the most critical steps in the process. Inadequacies and shortcomings in this step often result in rework and lost credibility. How many times have we found ourselves in receipt of report requirements via email, voice mail, or "back of the envelope" notes? Often we're all too eager to start our project on the basis of cryptic one-way communications from our customer. Taking the time, up front, to define the questions we will answer is a worthwhile investment. It provides us the opportunity to identify and document important considerations such as assumptions, and even the more mundane details such as report titles. But perhaps the best reason for using a systematic approach for defining the questions is the opportunity to shape client expectations by establishing a realistic understanding of the level of effort and complexity that need to be factored into the delivery time frame.

The 5 "W's" and the 1 "H"

When faced with a report requirement, there are five fundamental questions that when answered, provide the context for the work that will follow. Those classic questions, or the "5 W's" are:

- **Who** needs this information?
- **What** questions do they want answered?
- **When** do they need the answers?
- **Where** are the data?
- **Why** is this important?

Getting answers to these questions can provide you with valuable insights into what the requester really needs. Not all requesters can effectively articulate their needs, and sometimes what we are asked to deliver is not what needs to be delivered. Using the "5 W's" to get inside their heads can be a useful tool for shaping the requirements to deliver what the requester is seeking, even if they have difficulty specifying it.

Being able to answer the "1 H" question, i.e., "**How** did I get the answer?", is essential when presenting the results.

Output Requirements

So, how do we proceed? What's a good method for capturing this information? Our basic objective at this point in the process is to fully understand and document the output requirements, that when met, will satisfy the information needs of the requester.

Use a Template

I have found great value in working from a template of standard topics (see Figure 2.1). Using a template, or checklist, helps ensure that important issues won't be overlooked. It's a useful reminder of what needs to be covered in the course of gathering output requirements. It's like the checklist that an airline pilot uses before landing. For example, think of the adverse implications of skipping an important step like lowering the landing gear before touchdown.

Figure 2.1 - Output Requirements Template

Date:	Requester:		Analyst:	
Project:		Report Name:		Rpt. #
Business Purpose				
Question(s) to be Answered				

1. Definitions:

Inputs	
Assumptions	
Definitions	
Calculations	
Outputs	

2. Output Format:

Title	
Header	
Col. Headings	
Row Content	
Totals/Sub-Totals	
Footer	
Graphics	
Other	

Orientation:	☐ Portrait	☐ Landscape	Example Attached:	☐ Yes	☐ No

3. Output Logistics:

Medium	☐ Hardcopy	☐ Data	☐ HTML	☐ Other
Frequency	☐ Daily	☐ Monthly	☐ Quarterly	☐ Other
Type	☐ Scheduled	☐ Ad Hoc	☐ User Run	☐ Other
Delivery Location:				
Security/Access Issues:				

4. Agreement to Proceed:

Estimated Effort:	Due Date:
Requester Signature:	Date:

5. Change History:

Date	Requester	Change	Est. Effort	Due Date

Acceptance: _____   Date: _____

I know that some analysts have an aversion to using "forms". They feel constrained as if being forced to "paint by the numbers" rather than working from a blank canvas. However, before you dismiss this approach, consider the advantages and efficiencies of using a template as your starting point. This is particularly true for new analysts or for dealing with difficult customers. Having a set of topics that "must be completed" can be a useful tool for engaging such individuals and convincing them to spend adequate time with you to document their requirements. It can facilitate a meaningful meeting that otherwise might not occur.

Finally, don't feel that every blank must be filled in. Use the relevant portions of the form and skip the rest. Nor is the template intended to cover every possible issue. Add to it as needed.

Header Block

The first block in the Output Requirements template captures important information about the context and business purpose of the task. It addresses the **Who**, **What**, and **Why** of the "5 W's". Key players are identified by name and the specific business questions to be answered are listed. The business importance of these questions should also be noted. It's essential to actually state these needs as questions.

Definitions Block

The Definitions block is intended to explicitly document the tacit information that is typically residing in the requester's head. Achieving clarity in these areas is an important prerequisite to the data manipulation that will follow.

Inputs

Some requesters have an excellent grasp of the data needed to answer their questions; others may be clueless. Try to determine as much as you can about the number of data inputs, their possible source, likely data quality, and information about key fields and relationships between fields. Also determine whether you will be sampling or using entire ranges of data. Be prepared to discuss these issues with the requester in layperson's terms rather than technical jargon.

Assumptions

Requesters may have certain expectations about the data that they assume to be true. Your job is to ferret out these assumptions, document them, and then explore the data to see if they are valid. For example, you may encounter assumptions such as:

- "All claims incurred during CY 2002 were paid by January 31, 2003."
- "Age is always calculated by subtracting the birth year from the current year."
- "Every employee in the data set has a value for annual salary."

Assumptions must be identified, tested, and documented. Results must always be interpreted in light of any assumptions.

Definitions

Take the time to get a precise definition for all key variables. For example, when dealing with categorical fields such as "Payment Type", "Referral Source", or "Color", we need to obtain a listing of all possible values that could appear in the data set. The data dictionary (if it exists) is very helpful in this regard. In other cases we need to gain insight into how key terms are defined. For example, what does the requester mean by "fiscal year", "resource type", or "employee overhead loading factor"? Discussion of these terms might lead to the following clarifications:

- Fiscal Year = October 1 through September 30.
- Resource Type = This is a field in the data set that distinguishes employees from temporary labor. Employees have a value of "E" and all others have a value of "C".
- Employee Overhead Loading Factor = the multiplier applied to an employee's base salary to estimate the value of benefits and other headcount-based fixed costs. For 2004 the value is 1.32. For 2003 the value is 1.31.

Definitional differences can be identified and documented. For example, in the third illustration above, it may be important to know that the loading factor varies between years.

<u>Calculations</u>

In situations where a calculation will be required, work through the development of the actual formula with the requester. This will help ensure that additional assumptions are not overlooked. For example, if "age" has been defined as the difference between birth date and 12/31/03, then one possible formula would be:

$$age = int((mdy(12,31,03) - birthdate)/365.25);$$

Subtle but significant differences can exist between alternative formulas. Don't wait until you're presenting your results to find out that you and the requester had different views on how to calculate key variables.

<u>Outputs</u>

When the requester says "all I want is a report", don't assume anything. The term report is much too general. Start probing into the specifics. This will give you a jump-start on the next two sections of the template. Is the requester thinking of a hard copy document, and if so, will it include color? What about reporting via HTML or CD? Does the data set need to be available for subsequent analysis? If so, then it becomes a required output as well. Will the output be distributed, and if so, how?

In the course of this discussion you can help the requester think of alternatives to traditional paper documents.

Output Format

In this section of the template you can get very specific about the format details for the desired output. Even though the items in this section seem obvious, if left unspecified, expect a lot of modifications to your output after the requester sees the initial document.

Mocking up an example of what the report will look like, to include headings, spacing, orientation, and all the other attributes, can help the requester visualize the final product. Because "a picture is worth a thousand words", you may find yourself making design changes at this point, which is much more efficient than doing so after you've coded the report.

Output Logistics

This section contains a set of production-oriented issues that can have a significant impact on your development effort. Capturing these issues early is essential for accurately estimating when the final product can be delivered. For example, if you learn that access to the output must be restricted, you need to incorporate that requirement into your design plan and work plan estimates.

Agreement to Proceed

In this section of the template, you distill everything you've learned from the previous sections into an overall estimate of effort required to deliver the output by a specified date. When you reach agreement with the requester regarding the output requirements and delivery date, you should request a signature. Some analysts are uncomfortable with this formality and feel it runs against the collegial relationship they are trying to maintain. There is some truth to that perception. It does make the exchange seem more like a business transaction between buyer and seller than a handshake agreement between colleagues. However, I have found that obtaining a signature helps to ensure that the requester has given serious and adequate thought to what they're asking you to do. It also generates a higher level of commitment from the requester to what has been specified in the document.

Managing Change

In spite of the thoroughness with which we attempt to capture report requirements, there is still the potential for change. Once again, analysis is iterative. We tend to understand more about what we really want after viewing some initial results. Therefore, the goal for the analyst is not to avoid value added change, but to manage it. The final section of the template should contain a record of each requested modification. If such changes add to the level of effort or impact the due date, it should be noted in this section and communicated to the requester. The value of managing expectations cannot be overstated. Be clear and explicit when asked to change project scope that has been previously agreed to. The requester may have second thoughts if the impacts of those changes to the project cost or delivery date are incompatible with other priorities.

Managing change is difficult. Some analysts would rather absorb the cost of change out of their own time than negotiate with the requester. That's a decision each of us has to make. If you are unwilling or unable to work countless hours, find a way to prevent "scope creep" from taking over your life.

Achieving Final Acceptance

All good projects come to a successful conclusion, or so we hope. How do you know when you are really done? How do you know if the requester is satisfied with your work? The value of a signature comes into play again. When you've presented the final results and it appears that you've met the requester's requirements, ask for a signature at the bottom of the Output Requirements document. If the requester is truly satisfied with your work, there will be no hesitancy. Otherwise, you have a pretty good indication that your work is not yet finished.

Summary - Key Points

- Open your mind to the value of using a template for capturing requirements.
- Assumptions are like land mines. Find out where they are before you step on one.
- Managing change is difficult. Control it or it will control you.
- A signature from the requester is a powerful indicator of commitment and acceptance.

3. Getting Organized

Introduction

One of the hardest things to do at the beginning of a project is to take the time to gather your thoughts, think about how the project will unfold, and set yourself up for success by organizing what you're going to do. In the heat of battle, we're anxious to move forward toward the objective in the spirit of "ready-fire-aim". It takes considerable discipline to get organized before you start to write code, i.e., "ready-aim-fire". The adage "You can pay me now or you can pay me later, but you're going to pay me" applies at this stage of an analysis project.

This chapter will discuss several methods and standards that should be established before coding begins. Adherence to these standards will be the focus of Chapter 8.

Programming Standards

Every program should contain some basic identifying information as well as documentation of key items that are associated with the program. Writing code with a consistent "look and feel" can contribute to the clarity of the statements and ease of maintenance over time. The use of program and output naming conventions is a great technique for linking related items. Remember that as you move forward with an analysis project, you are actually creating pieces of a puzzle. The number of puzzle pieces is a function of the complexity and size of the project. At the end of the project, it's nice to be able to put the puzzle together. It's also nice to be able to reassemble the puzzle if were to inadvertently fall from the table to the floor into a jumble.

Naming Conventions

Program, variable, and file naming conventions provide a systematic method for saving each component of a project. These conventions are interpretable links that associate related pieces of work. Their use can dramatically simplify project documentation and your ability to reconstruct events, tasks which are often complicated as components are passed from one analyst to another, or simply by the passage of time.

<u>Program Names</u>

A program naming convention should connect the program to a project and programmer, provide some information about the type of processing that occurs in the program, indicate execution sequence, and provide version control. The convention illustrated below consists of an eight-character code that uniquely identifies each program. SAS programs use the .sas extension.

Format	Project Code	Programmer Initials	Task Type	Sequence Number	Version Number
Proj. Code	Assigned by the Project Manager				
Programmer Initials	two characters				
Task Type	R - Reading V - Validating C - Combining D - Describing M - Manipulating T - Testing P - Reporting X - Other				

Examples	RUDBR01A =	RU	=	Resource Utilization Project
		DB	=	D. Blakeley
		R	=	read data
		01	=	first program in the sequence
		A	=	version A of the program
	HCDBP02C =	HC	=	Health Care Claims Analysis Project
		DB	=	D. Blakeley
		P	=	reporting
		02	=	second program in the sequence
		C	=	version C of the program

Variable Names

Variable names should be descriptive, not cryptic. Take advantage of the length limitations of your operating system and programming language to define meaningful names, although not unnecessarily long. You may want to define names that associate a variable with its source, or with other variables in a group, or with the type of data that the variable represents.

SAS Files

Where possible, file naming should conform to the convention of associating items with the program that calls them or outputs them. In naming output files, use the name of the originating program as the first part of the output file name. For log and report files, suggested file extension conventions are shown below. In cases where there are multiple data set outputs, the program name should be appended with additional information that distinguishes the data sets. Note that the SAS System automatically assigns file names to user defined formats and the file extension for output data sets.

Type	File Name	File Extension
Data Set	same as program name	.sas7bdat
Log	same as program name	.log
Report	same as program name	.lst
Format	formats	.sas7bcat
Macro	same as the calling program name	.mac

For example, if a program named RUDBR01A.sas creates two output data sets and one report, the output file names could be defined as:

File Type	File Name
Data set one	RUDBR01A_ONE.sas7bdat
Data set two	RUDBR01A_TWO.sas7bdat
Report	RUDBR01A.lst
Log	RUDBR01A.log

Program Structure

Programs that are readable and interpretable are much easier to work with. As they become less like "code" and more like a logical narrative that documents what you're doing with your data, you'll find that the need for additional external documentation is reduced. A related advantage is that your programs can be more easily understood and used by other analysts.

Program Template

I strongly advocate creating a program template that becomes the starting point for every program you write. The first step in writing a new program is to open the template in the editor and assign a program file name based on your naming convention. Other elements in the comment block, such as source and programmer name, can be filled in immediately. Complete the remaining elements as the information materializes.

```
*-----------------------------------------------------------------------*
|                                                                       |
|Program:      program name                                             |
|                                                                       |
|Source:       location of program                                     |
|                                                                       |
|Purpose:      brief description of program purpose                     |
|                                                                       |
|Input:        input file names                                         |
|                                                                       |
|Output:       output file names and type, e.g., file, print           |
|                                                                       |
|Update Log:   Notes                           Date         Programmer  |
|              ----------------------------    -----------  ----------- |
|              Original version                mm/dd/yyyy    complete name|
|                                                                       |
|Usage Notes: e.g., identify macro parameters                           |
|                                                                       |
|Other Notes: e.g., any other information useful to the individual who will|
|             run the program or interpret the output                   |
|                                                                       |
*-----------------------------------------------------------------------*;
```

Over time, the comment block becomes a valuable part of your project documentation, as shown in the example below.

```
*-----------------------------------------------------------------------*
|                                                                       |
|Program:      RUDBR01C.sas                                             |
|                                                                       |
|Source:       c:\Projects\IT Resource Utilization\code\prod            |
|                                                                       |
|Purpose:      Reads the Lotus Notes extract and outputs a permanent SAS data|
|              set and a sample listing of the first 20 records.        |
|                                                                       |
|Input:        c:\work\LNdata\timedata                                  |
|                                                                       |
|Output:       RUDBR01C.sas7bdat                                        |
|              sample print to output window                            |
|                                                                       |
|Update Log:   Notes                           Date         Programmer  |
|              ----------------------------    -----------  ----------- |
|              Original version                05/12/2003    D. Blakeley |
|              Added PAD option                05/12/2003    D. Blakeley |
|              Added logic to read Hours       05/12/2003    D. Blakeley |
|              beginning in either col 58 or                            |
|              col 78                                                   |
|                                                                       |
|Usage Notes: The pad option is used with the infile statement because  |
|             there is record length variability within record type.    |
|             Check the SAS Log to ensure that the sum of               |
|             countblank + countname + countdate + countzero + output   |
|             data set records = total records read.                    |
|                                                                       |
|Other Notes: The input file is a Lotus Notes view of Period/Name/Project|
|             that was exported from Lotus Notes as tabular text.       |
|                                                                       |
*-----------------------------------------------------------------------*;
```

Writing SAS Statements

The SAS language allows considerable flexibility in the way SAS statements are written in a program. This level of flexibility means that the same code can be entered into a program in many different styles and formats. While the code would certainly compile and execute identically, the ability to read, understand, maintain, and modify the code can vary widely, based on the style with which it is entered. If programmers follow a standard, it is more likely that code can be easily shared and supported across programmers.

The statements in a SAS program are divided into two kinds of steps: DATA steps and PROC steps, the building blocks of most SAS programs. The DATA step can include statements asking SAS to create one or more new SAS data sets and the programming statements that perform the manipulations necessary to build the data sets. Report writing, file management, and information retrieval are all handled in DATA steps. The PROC step asks SAS to call a procedure from its library and to execute that procedure, usually with a SAS data set as input.

SAS statements can begin in any column of a line, and several statements can be written on the same line. At least one blank between each separate item in a SAS statement is required. Some special characters, such as the equal sign after a word, can take the place of a blank, although blanks are always allowed. For example, the following two statements are equivalent:

ASSETS=LIABILITY+EQUITY;

ASSETS = LIABILITY + EQUITY;

Although SAS does not have rigid spacing requirements, SAS programs are easier to read if the statements are indented consistently, based on the following spacing conventions:

- Indent secondary statements and continuations of statements within a DATA or PROC step.
- Clearly separate steps by inserting a blank line.
- List variables vertically rather than horizontally.
- Organize IF/THEN/ELSE or other conditional statements into logical groupings.

The two SAS programs displayed below are equivalent. Which one is easier to read?

```
title1 'Global Title';

data two;
   infile 'c:\projects\projectABC\rawdata';
   length admit $8.;
   input    @ 1  var_one $10.
            @12  var_two  2.
            @14  varthree 2.
            @16  var_four $9.;
   * distinguish between emergent and non-emergent admissions ;
   if var_two='xx' then do;
      admit='emergent';
      end;
   else
      admit='non_er';
      drop var_two;
run;

title2 'Date Set Two';
proc print data=two;
   var var_one
         admit
         varthree
         var_four;
run;
```

```
title1 'Global Title'; data two;
infile 'c:\projects\projectABC\rawdata'; length admit $8.;
input @ 1  var_one $10. @12  var_two  2. @14  varthree 2.   @16
var_four $9.;
* distinguish between emergent and non-emergent admissions ; if
var_two='xx' then do;
admit='emergent'; end; else
admit='non_er'; drop var_two;
run; title2 'Date Set Two'; proc print data=two;
var var_ona admit varthree
var_four; run;
```

Code Comments

Programmers seem to be divided into two camps: Those that comment their code and those that don't. Comment statements strategically inserted into a program provide an important source of project documentation. The information contained in a comment statement can illuminate and describe a section of code that might otherwise be cryptic, ambiguous, or confusing. A variable transformation or conditional statement might make perfect sense at the moment you write it. But when revisiting that same uncommented code several months later, the effort required to interpret that code and the context in which it was created could be extensive.

Comment statements are simple to use and powerful when used consistently and appropriately. A comment statement is a section of text inserted into a program and then ignored by the program. One method for inserting a comment into a SAS program is to begin the comment with an asterisk (*) and end it with a semicolon. Enter as many lines of text as necessary between the asterisk and semicolon. The sample program in the section above shows one example of a comment statement. Comments should be located near the statements they describe and can also appear on the same line as other SAS statements, as in the example below.

```
CCRATIO  = AVGCST/AVGCHG; * calculate the cost-to-charge ratio ;
```

Use comments to describe why you coded something in a particular way, why a statement option was used, or to clarify the logic behind a complex piece of code. There should be no section of code where there is any ambiguity about what is being done or why it is being done. However, guard against "over-commenting", i.e., commenting the obvious as in the example below.

```
* Print the report ;
proc print;
run;
```

Finally, I encourage you to document code as you write it. This habit does not come naturally. It's a practice that most programmers must learn and then consistently apply. The longer you wait to add comments, the less likely they are to be adequate, accurate, meaningful, or done at all.

Organizing Your Materials

Data analysis is inherently iterative involving cycles of exploration, code testing, production runs, and so forth. Then just when you think you're done, the research questions change, and you start all over. This can be further complicated when you are working on multiple projects at the same time. In order to avoid chaos, it's imperative that an appropriate level of structure and organization be applied to this activity. One such scheme is outlined below.

Folders

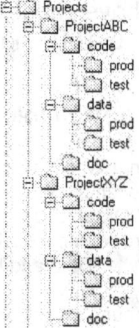

This is a simple framework for organizing the basic components of most projects. Set up folders or other organizing mechanisms supported by your operating system that allow you to separate code from data, test versions from production versions, and provide a repository for project documentation. Some projects may require additional folders, e.g., to hold a format catalog. Also note that a separate folder structure should be used for each project. Adhering to a framework such as this can help avoid mistakes.

Libraries

The SAS System offers a structure for organizing and storing SAS data sets and other SAS files. Every SAS file is stored in a SAS data library, which has different implementations depending on your operating environment. In our case, a library is simply a group of SAS files in the same folder or directory. To reference a permanent SAS data set, you use a two-level name:

> *libref.filename*

In the two-level name, *libref* is the name of the SAS data library that contains the file, and *filename* is the name of the file itself. A period separates the libref and filename. For example, SASIN.RATES is the two-level name for the SAS data set RATES, which is stored in the library named SASIN.   Use a LIBNAME statement to define a libref and associate it with a specific library. This example defines the libref SASIN.

```
        LIBNAME SASIN 'c:\Projects\ProjectABC\data\prod';
```

You can see that the library is simply a path to a physical location. To read or write a permanent SAS data set, you specify both the first- and second-level names in the SAS statements referring to the file. The libref points to the location where the file is (to be) stored. For example, when these statements execute

```
        LIBNAME SASOUT 'c:\Projects\ProjectABC\data\prod';
        DATA SASOUT.PRODUCTS;
          additional statements ...
        RUN;
```

SAS writes a file called PRODUCTS in the storage location referenced by SASOUT. To use this file in a subsequent step or another program, specify the appropriate libref followed by the second-level name. For example:

```
        PROC PRINT DATA=SASOUT.PRODUCTS;
```

Data Management Concepts

"Data Management" means different things to different people. For some it refers to the technical aspects of administering databases. For others it relates to the process of capturing, classifying, and storing data for scientific purposes. It can also focus on the documentation of data quality and completeness. For the purposes of this book, I have chosen to define data management in terms of the set of principles or axioms, as set forth in chapter 1, and reinforced again below.

Guiding Principles

If It Isn't Written Down It Was Never Said Or Done
Identify and Document Assumptions
Get Organized
Look For Problems Early
Seek Out Master Data and Use It
Stay Organized
Archive Data As If Your Career Depended On It
Documentation Should Pass the "Hit By a Bus" Test
Expect To Be Questioned -- Possibly Doubted
Analysis Is Iterative
Confidence Comes From Knowing You Can Replicate Your Results

Summary - Key Points

- Get organized <u>before</u> you start coding.
- Whatever standards you use, use them <u>consistently</u>.
- Document code as you write it, not after.
- Remember that data analysis is iterative.

4. Gathering the Data

Introduction

Truly understanding the questions to be answered positions us to obtain and process the appropriate data. Achieving the gold standard of "Getting it right the first time" is absolutely dependent on having the proper data. This step focuses on the logistics of getting data quickly and efficiently. Identifying who to contact, the options that may exist regarding format and delivery medium, how to specify your request, and jointly agreeing on when the data will be available, can significantly lessen the amount time devoted to this stage of the project.

This chapter will focus on the effective use of the "Input Requirements" template (see Figure 4.1).

Input Requirements

The first block in the Input Requirements template captures basic information that links the document to the corresponding Output Requirements that were discussed in Chapter 2. It's quite possible that many of the categories in this template can be completed at the same time the Output Requirements are being defined. It usually depends on the technical depth of the requester and how well he or she knows the data and data sources.

Contact Information

After completion, the Input Requirements form is typically emailed or otherwise delivered to either the data source business contact (section 2) or the data source technical contact (section 3).

Requester Contact

Provide compete contact information for whomever is requesting the data, which is many cases may be you, the analyst. The context for section 1 of the template is the data requester, not necessarily the requester associated with the Output Requirements document.

Source Contacts

Depending on the complexity of the data and the knowledge of the source contacts, you may need multiple contacts for addressing questions you have about the source data. Having both a business or functional area contact as well as a technical contact gives you more options for dealing with tough questions.

Data Requirements

Section 4 of the template is very important. This is where you specify what you would like to receive. Based on your knowledge of what the data source can provide, you may have to consider tradeoffs between the ease with which you will be able to process the data versus the level of effort required of the data source to prepare the data to your specifications. For example, if it's faster and less expensive for the data source to send you a hierarchical file on tape than to reformat it into fixed length records on a CD, you may prefer to get the data sooner and spend your time reformatting it.

Content

Be very specific about data time periods or other attributes that determine the record selection criteria that the data source will apply to the extract they prepare. Leave nothing to assumption. If possible,

Figure 4.1 - Input Requirements Template

Date:	Project:			
Nature of Request				
Date Required:		Rush:	☐ Yes	☐ No

1. Requester Contact:

Name	
Organization	
Postal Address	
E-mail Address	
Telephone:	Fax:

2. Source Business Contact:

Name	
Organization	
Postal Address	
E-mail Address	
Telephone:	Fax:

3. Source Technical Contact:

Name	
Organization	
Postal Address	
E-mail Address	
Telephone:	Fax:

4. Data Content/Format:

File Type: ☐ Fixed Field ☐ Free-Format ☐ Hierarchical ☐ Other
Delimiter info.
Time Period
Data Volume   Size: Records:
Documentation: ☐ Record Layout ☐ Data Model ☐ Data Dictionary ☐ Other
Medium: ☐ Tape ☐ CD ☐ Hardcopy ☐ Other

5. Send To:

Name	
Organization	
Delivery Address	

6. Ship Via:

U.S. Mail	☐ First Class	☐ Parcel Post	
UPS	☐ Next Day	☐ 2nd Day	☐ Ground
FedEx	☐ Next Day	☐ 2nd Day	☐ Saturday Delivery
Other			

7. Receipt Information:

Date Received		Received By	
Date Loaded		Loaded By	
Filename(s)		File Location(s)	
Date Reviewed		Reviewed By	

try to get an understanding of how large your extract will be. Will you have enough storage space to accommodate it?

Format

If you have format options, be as specific as possible about how you prefer the data formatted.

Documentation

Request all the documentation that you think you will need, such as data dictionaries, record layouts, data flow diagrams, data models, etc. The need for repeated requests for this type of information is not unusual.

Shipping Requirements

Be as specific as possible about how you want the data shipped, when it needs to arrive, and to whom it should be sent. This avoids the potential for shipments to arrive at the mailroom and sit for days while the mail clerk tries to figure out where the package belongs. It also gives you an indicator of when to start looking and what to look for in the event the shipment doesn't arrive when expected.

Tracking

Keep track of when the data arrives and what you did with it. If others need to be notified and/or review the data, make note of those events in section 7 of the template.

Summary - Key Points

- A lot of valuable project time can be lost waiting for data.
- The only thing worse than data arriving late, is when it arrives late and incorrect, due to poor specifications.
- Be very specific about what you want -- you never know what you can get until you ask for it.
- Keep track of all shipping/receiving events. It's useful information the next time around.

5. Creating the Analysis Data Set(s)

Introduction

This step typically involves a great deal of data reading, manipulation, and output. It is the Extract-Transform-Load part of the project. As such it presents plenty of opportunities for error, particularly in relation to any assumptions we still have regarding the data. The analysis dataset(s) created during this step form the foundation for everything that will be delivered from the project.

When discussing data, multiple terms are often used interchangeably to refer to the same concept.

Concept	Referred To As
Groupings of data	Data set, file, database
Columns of data	Field, variable, data element
Rows of data	Record, observation

Reading the Data

Once you've received the data and associated documentation, the next step is to begin the incremental process of reading and understanding the data. This is more of an evolutionary process than a big bang event, particularly if the data files are large or complex.

Remember to use naming conventions for program, variable, and file names as discussed in Chapter 3.

Data Types

There are two data types that you will work with: numeric and character. Every variable you create will be either numeric or character.

Values for standard numeric data can contain only:

- numbers
- decimal points
- numbers in scientific or E-notation
- plus or minus signs.

Nonstandard numeric data includes:

- values that contain special characters, such as percent signs (%), dollar signs ($), and commas (,)
- date and time values
- data in fraction, integer binary, real binary, and hexadecimal forms.

The character data type contains data values that are treated as a succession of characters.

First Glimpse

Reading unfamiliar data requires attention to detail. It's not inherently difficult, but very prone to "small" errors such as specifying the wrong input format or incorrectly stating the position of the fields you desire to read. Therefore, it's often advisable to start conservatively using the following guidelines:

- Print out one entire record to view what the data actually look like. Compare the physical characteristics of the data to what the documentation indicates should be there. Be alert for any inconsistencies. For example, if you were reading a file with fields for ID, age, gender, height in inches, and weight in pounds, the following code would read the first observation and produce the SAS log shown beneath it. Note that the SAS log shows a value of 395482753 for ID, 49 for age, M for gender, 72 for height, and 185 for weight. All values appear to be valid, at least for this observation.

```
data _null_;
  infile 'c:\Projects\Height-Weight\data\prod\Height-Weight.txt' obs=1;
  input;
  put _infile_;
run;

-------------------------------------------------------------------------------
NOTE: The infile 'c:\Projects\Height-Weight\data\prod\Height-Weight.txt' is:
      File Name=c:\Projects\Height-Weight\data\prod\Height-Weight.txt,
      RECFM=V,LRECL=256

395482753 49 M 72 185
NOTE: 1 record was read from the infile 'c:\Projects\Height-
Weight\data\prod\Height-Weight.txt'.
      The minimum record length was 21.
      The maximum record length was 21.
NOTE: DATA statement used:
      real time          0.46 seconds
      cpu time           0.04 seconds
```

- Save coding and processing time by reading only those fields and records that you need.
- Before attempting to read the entire file, start by reading only a few records until you are sure that you are accurately processing the data.

As you're testing your code, increase the number of observations read with each successive pass through the data. As you continue to move deeper into the file, don't be surprised if you encounter unexpected data anomalies that didn't surface in the previous records.

Data Conversion

In addition to reading raw data as described above, there will be other occasions where your input data are in the form of vendor-specific database products such as Oracle®, IBM® DB2®, and other relational database management systems (DBMS). Microsoft® Access and Excel files are also common external data sources. The SAS System provides SAS/ACCESS® software which enables SAS to share data with a number of DBMSs where the data can be accessed and referenced in your SAS programs as if they were in SAS data sets. These conversion engines are very powerful and a tremendous time saver.

The SAS System also provides other mechanisms, such as the IMPORT procedure, for reading data from external PC files and writing it to a SAS data set. This makes the processing of Access and Excel files a snap.

Sampling

Another consideration when reading the data is whether a sample will suffice. In some cases the complete file may be too large to work with comfortably, which makes sampling a more appropriate approach. When sampling, keep the following considerations in mind:

- If the order of observations is random then use any arbitrary rule to select a certain number of observations.

- If the observations are not in random order then you will need to randomize them, most likely by sorting by a variable whose value is a random number.
- Will you select a sample of an <u>approximate</u> size or an <u>exact</u> size?
- Are you sampling with replacement?
- Are you creating a grouped or stratified sample?
- Systematic sampling is an approach that selects observations by a mathematical process that is not random. Periodic sampling is an example of systematic sampling, such as "Select every 5th observation".

All of the sampling issues listed above are easily implemented using a variety of SAS capabilities.

Conditional Selection

In situations where you want to use only a portion of the observations available to you and disregard the others, conditional selection, or subsetting, is appropriate. Sampling is certainly a form of conditional selection, although it is constrained by the need for the subset to represent the larger set in some manner. This is particularly important for statistical analysis. When your need is to focus on a certain non-random portion of the data, such as a specific type of transaction or all events within a date range, then subsetting techniques such as those listed below can be used:

- Use the WHERE statement to select observations that meet a specific condition. For example, WHERE PRODUCT = "shirt" AND COLOR = "red";.
- Use the WHERE data set option to select observations from a SAS data set that meet a specific condition.
- Use a subsetting IF statement in a SAS data step. The data step stops processing any observations that do not meet the specified condition. It continues processing observations that meet the condition. For example, IF PRODUCT = "shirt" AND COLOR = "red";.
- The DELETE statement used in conjunction with an IF-THEN statement has the same effect as a subsetting IF statement when the inverse condition is specified. For example, IF PRODUCT NE "shirt" AND COLOR NE "red" THEN DELETE;.
- Use the OUTPUT statement in conjunction with an IF-THEN statement to write selected observations to an output SAS data set. For example, IF PRODUCT = "shirt" AND COLOR = "red" THEN OUTPUT;.

The @ sign used at the end of an INPUT statement (a.k.a. "trailing at sign") is another method of conditional selection that can be used during input processing. The trailing at sign (@) holds the input record for the execution of the next INPUT statement or when control returns to the top of the DATA step. This is a very efficient method for screening observations for a specific condition, such as STATUS = "paid". Only observations with a value of "paid" will be processed. All other observations will be deleted.

Validating the Raw Data

This section begins our discussion of issues related to data completeness, which will be continued in greater detail in chapter 6. Consider this the initial check of data completeness intended to answer the question "Did we get the data we requested?" In other words, the data completeness test is a measure of convergent validity, i.e., the degree to which independently constructed measures agree.

We are looking for evidence that all expected records were <u>received</u> and that all required fields are <u>complete</u>. In order to determine these measures we need some form of external comparison. Hopefully the organization that produced the data is able to provide control reports, data dictionaries and other documentation that will be a useful comparison to the results we obtain. External comparisons are often in the form of records counts, numeric totals, time period ranges and other descriptors of content that you will be able to replicate, if you received all the data you were expecting. Several SAS procedures will be discussed in chapter 6 that are ideally suited for the production of quick and easy comparative reports.

Reviewing the Results

After each pass through the data, be sure to review the SAS log, paying attention to notes and any errors that are listed. The SAS log is an extremely useful tool for identifying and diagnosing data problems. In the log displayed below, we see several important pieces of information.

```
NOTE: The infile 'c:\Projects\Height-Weight\data\prod\Height-Weight1.txt' is:
      File Name=c:\Projects\Height-Weight\data\prod\Height-Weight1.txt,
      RECFM=V,LRECL=256

NOTE: Invalid data for ID in line 1 1-9.
NOTE: Invalid data for weight in line 1 19-21.
RULE:       - - - +- - - -1- - - +- - - -2- - - +- - - -3- - - +- - - -4- - - +- - - -5- - - +- - - -6- - - +- - - -7- - - +- - - -8- - - +-
1          A95482753  49  M  72  A85  21
ID=. age=49 gender=M height=72 weight=. _ERROR_=1 _N_=1
NOTE: 13 records were read from the infile
      'c:\Projects\Height-Weight\data\prod\Height-Weight1.txt'.
      The minimum record length was 21.
      The maximum record length was 22.
NOTE: The data set WORK.CH5EX has 13 observations and 5 variables.
NOTE: DATA statement used:
      real time  0.07 seconds
      cpu time   0.05 seconds
```

- The first note names the file that was read.
- The second and third notes indicate that invalid data were found in the fields ID and weight. The observation containing the invalid values is listed beneath the ruler. You can see that the values for ID and weight each contain a character, which is incompatible with the expected numeric format. This is likely the result of data entry error. For this observation, the values for ID and weight are set to missing.
- The fourth note indicates that 13 records were read.
- The final notes states that the output data set contains 13 observations and 5 variables.

Make a thorough review of the SAS log a standard part of your process.

Printing

If the messages in the SAS log indicate that you are correctly reading the raw data file, it is a good idea to print several observations before reading the entire raw data file. This sample or test printing is another opportunity to gain insight into possible data problems. The code and output below demonstrate this technique and a listing of the first five observations that were read from the raw file and written to a SAS data set named "ch5ex".

```
    title 'Sample Listing';
    proc print data=ch5ex;
    run;

    ------------------------------------------------
                        Sample Listing

    Obs      ID        age    gender    height    weight

     1    395482753     49      M         72       185
     2    695584945     40      F         70       150
     3    492203991     23      M         73       180
     4    600498593     42      M         71       190
     5    590938275     25      F         65       115
```

Counting Input and Output Records

Whenever using conditional logic to select observations, I recommend using a method to account for each record processed, even those that are deleted. This type of before-and-after comparison is used to ensure that you haven't inadvertently skipped or deleted relevant data. In the simple example that follows, we are reading a SAS data set called ch5ex that contains thirteen observations. We are only interesting in keeping the observations for females (gender = "F") weighing greater than 100 pounds (weight > 100). This is accomplished using a subsetting IF statement as shown in line 705 of the SAS log. In order to account for the observations that will be deleted, three additional statements are used to identify and count the males (gender = "M") and anyone with a weight less than or equal to 100 pounds (weight <= 100). These counts are assigned to the variables *countmales* and *countlightweights*, whose values at the end of the data step are written to the SAS log.

```
702  * keep only females greater than 100 pounds ;
703  data subset;
704     set ch5ex end=last;
705     if gender = "F" and weight > 100 then output;
706     else
707     if gender = "M" then countmales + 1;
708     else
709     if weight <= 100 then countlightweights + 1;
710     if last then put countmales= countlightweights= ;
711  run;

countmales=7 countlightweights=1
NOTE: There were 13 observations read from the data set WORK.CH5EX.
NOTE: The data set WORK.SUBSET has 5 observations and 7 variables.
NOTE: DATA statement used:
      real time     0.09 seconds
      cpu time      0.04 seconds
```

It is now easy to account for how each of the thirteen observations in the original data set were processed.

Observations read	13
Males deleted	7
Lt. weights deleted	1
Observations output	5

Observations output (5) = Observations read (13) - Males deleted (7) - Lt. weights deleted (1).

Combining the Data

An important part of creating the analysis data set is to anticipate the data structures that will be needed to answer the questions that have been defined. The ease with which the questions can be answered can be impacted, positively or negatively, by the way the data are organized for analysis. A poor job of structuring the data means that additional manipulation will be required, often at great cost. Some analysis goals might not be achievable if key data have been left out of the analysis data set.

If you will be sharing the data or completely turning it over to another analyst or group, keep in mind the preferences and data manipulation skills of the end user as you plan the structure of the analysis data set. For example, many analysts prefer to work with a single consolidated data set of fixed record length where all they have to worry about is a standard set of identically formatted rows and columns. Others are willing to work with more complex data structures such as hierarchical files or multiple tables.

Several techniques using the SAS System are outlined below for combining and restructuring data to accommodate the needs of the analysis plan.

Concatenation

The process that appends the observations from one data set to another data set is concatenation. Conceptually, it looks like this.

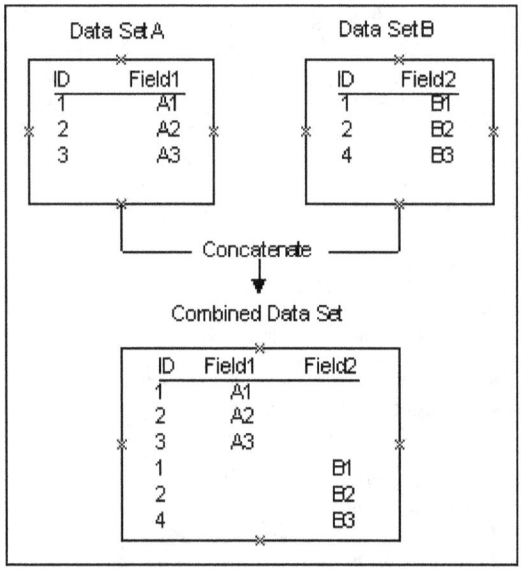

To concatenate SAS data sets, you specify a list of data set names in the SET statement, as in the following example.

```
data combine;
  set a b;
run;
```

All the observations are read from the first data set listed in the SET statement. Then all of the observations are read from the second data set listed, and so on, until all of the listed data sets have been read. The new data set contains all of the variables and observations from all of the input data sets. As you can see from this example, if one data set contains a field that is not in the other data set, the field will have a missing value in observations that are read from the other data set.

An alternative to concatenation is to use the APPEND procedure. In appending, observations are added to the <u>base</u> data set and observations of a different data set are added to the end of the base data set. This can be much faster than concatenating with a SET statement if the number of new observations is small in comparison to the number of base observations, because it is not necessary to read or rewrite any of the data in the base file. Consider using the APPEND procedure in cases where the two data sets have the same fields and your objective is to simply add new data to the end of the base data set.

Interleaving

If you use a BY statement when you concatenate data sets, the result is interleaving. Interleaving intersperses observations from two or more data sets, based on one or more common variables. Each input data set must be sorted or indexed in ascending order based on the BY variable. To

32

interleave SAS data sets, specify a list of data set names in the SET statement, and specify one or more BY variables in the BY statement, as in the following example.

```
data combine;
 set a b;
 by id;
run;
```

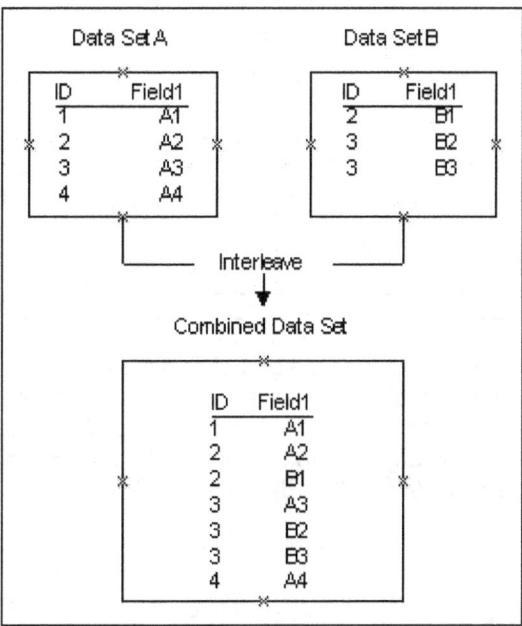

When the SAS System interleaves data sets, observations in each BY group in each data set in the SET statement are read sequentially, in the order in which the data set and BY variables are listed, until all observations have been processed. The new data set includes all the variables from all the input data sets, and it contains the total number of observations from all input data sets.

Match-Merging

The process for combining observations from two or more data sets into a single observation in a new data set according to the values of a common variable is called match-merging. Each input data set must be sorted or indexed in ascending order based on the BY variable. When you match-merge, you use a MERGE statement rather than a SET statement to combine data sets, as in the following example.

```
data combine;
 merge a b;
 by id;
run;
```

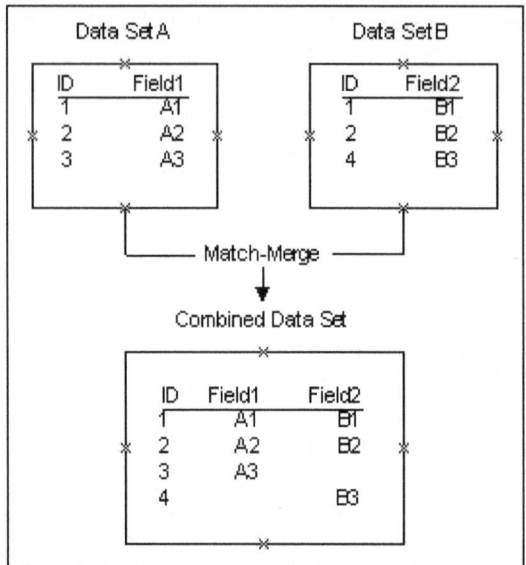

In general, the SAS System sequentially check each observation of each data set to see whether the BY values match, then writes the combined observation to the new data set. Simple DATA step match-merging produces an output data set that contains values from all observations in all input data sets. A typical use for match-merging is to combine data sets where one data set contains unique values for the BY variable and the other data set contains multiple values for the BY variable.

Data Transformations

As we have seen in this chapter, data can come in a variety of formats. The needs of your analysis may require you to transform, or reshape, the data set. This often takes the form of:

- creating a single observation from multiple records (changing observations into variables)
- creating multiple observations from a single record (changing variables into observations)
- collapsing hierarchical files
- flipping a SAS data set file on its side

Each of these techniques will be briefly reviewed.

Creating a Single Observation From Multiple Records

We might find a situation where we have name and address information spread across three records, as in the following example.

Raw Data File

```
SALLY REYNOLDS
183 CARRIAGE LANE
BAKERSFIELD CA 93301
TOM SMITH
2817 LILAC AVENUE
LONG BEACH CA 90745
MARY CARPENTER
44 OCEAN BLVD
MONTEREY CA 93940
```

If having one record per person better suits the analysis plan, we can use the forward slash (/) line pointer control to read multiple records sequentially. Using a single INPUT statement, we can read the value for first name (Fname) and last name (Lname) in the first record, followed by the values for street address (Address) in the second record. Then the values for city (City), state (St), and zip code (Zip) are read from the third record. The INPUT statement should be preceded with a LENGTH statement that defines Lname with a length of 9 to accommodate "Carpenter", which is the longest value for Lname. Also, because some of the values for City are longer than eight characters and contain embedded blanks, you should use modified list input with the ampersand (&) modifier to read these values.

The resulting SAS data set would look like this.

SAS Data Set

Obs	Lname	Fname	Address	City	St	Zip
1	REYNOLDS	SALLY	183 CARRIAGE LANE	BAKERSFIELD	CA	93301
2	SMITH	TOM	2817 LILAC AVENUE	LONG BEACH	CA	90745
3	CARPENTER	MARY	44 OCEAN BLVD	MONTEREY	CA	93940

Note that the Lname variable appears first in the data set because the LENGTH statement preceded the INPUT statement, causing the variable attributes for Lname to be defined before the other variables.

Creating Multiple Observations From a Single Record

Another situation you might encounter is a raw data file that contains data for several observations in one record. This is often done to conserve file space. In the following example we have a raw data file that contains three blocks of data. Each block contains a date followed by the total rainfall at a single location.

Raw Data File

```
01SEP03 000 02SEP03 .50 03SEP03 .10
04SEP03 000 05SEP03 1.0 06SEP03 000
07SEP03 1.1 08SEP03 000 09SEP03 .20
10SEP03 .12 11SEP03 000 12SEP03 000
13SEP03 .44 14SEP03 000 15SEP03 000
16SEP03 000 17SEP03 1.0 18SEP03 000
19SEP03 .90 20SEP03 000 21SEP03 000
22SEP03 000 23SEP03 .01 24SEP03 .03
25SEP03 .04 26SEP03 000 27SEP03 000
28SEP03 000 29SEP03 .56 30SEP03 000
```

By reformatting the data into a separate observation for each data block, it will be much easier to analyze the data with statistical procedures. Using the double trailing at-sign (@@) with a single INPUT statement, the INPUT statement will read the values for Date and Rainfall and execute three times for each record in the raw data file. A FORMAT statement is also included to display the values for Date in the form ddmmmyyyy. The resulting SAS data set would look like this.

SAS Data Set

Obs	Date	Rainfall
1	01SEP2003	0.00
2	02SEP2003	0.50
3	03SEP2003	0.10
4	04SEP2003	0.00
5	05SEP2003	1.00
6	06SEP2003	0.00
7	07SEP2003	1.10

```
8       08SEP2003       0.00
9       09SEP2003       0.20
10      10SEP2003       0.12
11      11SEP2003       0.00
12      12SEP2003       0.00
13      13SEP2003       0.44
14      14SEP2003       0.00
15      15SEP2003       0.00
16      16SEP2003       0.00
17      17SEP2003       1.00
18      18SEP2003       0.00
19      19SEP2003       0.90
20      20SEP2003       0.00
21      21SEP2003       0.00
22      22SEP2003       0.00
23      23SEP2003       0.01
24      24SEP2003       0.03
25      25SEP2003       0.04
26      26SEP2003       0.00
27      27SEP2003       0.00
28      28SEP2003       0.00
29      29SEP2003       0.56
30      30SEP2003       0.00
```

Collapsing Hierarchical Files

Your raw data files may have a hierarchical structure, consisting of a header record and one or more detail records. In most cases, each record contains a field that identifies the record type. In the example below, the 'C' indicates a header record that contains a customer's account number. The 'P' indicates a detail record that contains the date of a purchase and the charges incurred.

Raw Data File

```
C   1945882040069429
P   07-22-03   39.45
P   07-25-03   57.03
C   2906983045069483
P   07-23-03   125.87
C   3109654882340059
P   07-22-03   112.36
C   3509232845560390
P   07-24-03   25.96
P   07-27-03   259.45
C   7889423049965684
P   07-25-03   31.05
P   07-26-03   19.44
```

Depending on your analysis plan, you may decide to build a SAS data set from the hierarchical file by creating one observation per detail record and storing each header record as part of the observation.

SAS Data Set

```
Obs   Account            Date       Amount

1     1945882040069429   07-22-03   39.45
2     1945882040069429   07-25-03   57.03
3     2906983045069483   07-23-03   125.87
4     3109654882340059   07-22-03   112.36
5     3509232845560390   07-24-03   25.96
6     3509232845560390   07-27-03   259.45
7     7889423049965684   07-25-03   31.05
8     7889423049965684   07-26-03   19.44
```

You can also build a SAS data set from a hierarchical file by creating one observation per header record and combining the information from detail records into summary variables.

SAS Data Set

```
Obs  Account            Amount

1    1945882040069429   96.48
2    2906983045069483   125.87
3    3109654882340059   112.36
4    3509232845560390   285.41
5    7889423049965684   50.49
```

The SAS System provides all the tools necessary to conditionally process each observation, retain variable values and otherwise manipulate the data to achieve the desired outcome.

Flipping a SAS Data Set On Its Side

Suppose you have a SAS data set that you want to restructure so that selected variables become observations. The TRANSPOSE procedure will accomplish that task and create an output SAS data set that can be used in subsequent DATA or PROC steps for analysis, reporting, or further data manipulation.

The following example illustrates a simple transposition. In the input data set, each variable represents the grades for one student. In the output data set, each observation will represent all grades for one student.

Input SAS Data Set

Student1	Student2	Student3	Student4
B	A	C	B
A	A	B	B
C	P	A	I
A	B	C	A
A	A	C	C
I	B	B	F
C	A	D	A
B	A	A	B

Transposed Output SAS Data Set

NAME	COL1	COL2	COL3	COL4	COL5	COL6	COL7	COL8
Student1	B	A	C	A	A	I	C	B
Student2	A	A	P	B	A	B	A	A
Student3	C	B	A	C	C	B	D	A
Student4	B	B	I	A	C	F	A	B

Each value of '_NAME_' is the name of a variable in the input data set that the procedure transposed. Therefore, the value of '_NAME_' identifies the source of each observation in the output data set.

Programming Techniques

Data-Driven Programs

Many programs contain parameters that vary from one analysis to another, e.g., dates, identifiers, and other values. These parameters often represent only a small portion of a program. The traditional practice of "hard-coding" the parameter values creates programs that are limited to a specific analysis, even though most of the remaining code is "generic". When the parameter values change, the program must be modified before it can be reused. Modification can produce a variety of errors and can confound attempts to maintain an audit trail. An alternative to this approach is to develop programs that are "data-driven". These programs consist of "generic" code that rarely, if ever, changes. When these programs run, parameter values are read from external files that have been updated by you or another analyst. As a result, the basic program can remain fixed, with only the external data changing value.

SAS Macro Language Processing

The SAS System provides a very powerful tool for creating and using "data-driven" programs. The macro facility is a mechanism for extending and customizing the SAS System and for reducing the amount of programming effort required to repeat common tasks. The macro facility allows for the packaging of small or large amounts of code into named units. From that point on, you can work with the names rather than with the program.

The macro facility has two components:
- The macro processor is the portion of the SAS System that does the work. It simplifies repetitive data entry tasks and provides a way to store and retrieve SAS programs that must accommodate changing parameters.
- The macro language is used to communicate with the macro processor. It has the same kinds of capabilities found in SAS DATA step statements. However, the macro language used the percent sign (%) and ampersand (&) to trigger macro processor activity.

A macro is stored text containing SAS code and macro language statements that is referred to my name. When you invoke a macro, the macro facility generates SAS statements and commands as needed. The rest of the SAS System receives those statements and command as if they had been entered directly.

This section describes four techniques that use macro language features:

- Macro parameters
- Record selection
- Table look-up
- BY statement processing

The macro displayed below will be used to illustrate these techniques. Note that in this macro, comment statements begin with %* and end with a semi-colon.

```
%macro pricing(new,old,amount);

data &new;
  %* read paid invoice file ;
  set &old;
  %* read vendor list ;
  if %vendorlist;
  %* read product list ;
  if %prodlist then do;
    %* calculate simulated cost using preferred rate ;
    sim_cost = (wholesale * pre_rate);
    %* calculate simulated savings ;
    savings = (cost - sim_cost);
```

```
   end;
   else do;
     savings = 0;
   end;
   %* output if the savings is greater than or equal to the specified amount ;
   if savings >= &amount;
run;

proc sort data=&new out=report1;
   by vendor product;
run;

title1 "Listing of Selected Products From:  &new - Date: &sysdate";
title2 "Products Where the Simulated Difference Between";
title3 "Actual Cost and Simulated Cost is Greater Than or Equal to $&amount";
proc print data=report1 noobs split='*';
   var vendor
       name
       product
       wholesale
       cost
       sim_cost
       savings;
   by vendor;
   sum wholesale
       cost
       sim_cost
       savings;
   format wholesale
          cost
          sim_cost
          savings dollar7.;
   label vendor    = 'Preferred*Vendor'
         name      = 'Vendor*Name'
         product   = 'Product'
         wholesale = 'Wholesale*Price'
         cost      = 'Actual*Cost'
         sim_cost  = 'Simulated*Cost'
         savings   = 'Simulated*Savings';
run;

%mend pricing;
```

This macro is a simple simulation that uses a paid invoice data set to calculate what would have been paid to preferred vendors if they had offered a volume discount for certain products. The program selects vendors from the input data set if they are contained in the preferred list (%vendorlist). Only products contained in the product list (%prodlist) are repriced. A simulated savings is calculated as the difference between the actual cost (cost) and the simulated cost (sim_cost), where simulated cost is the product of the wholesale price (wholesale) and the preferred markup rate (pre_rate). Only re-priced invoices where the savings are greater than or equal to a specified amount (&amount) are contained in the output data set and subsequently printed in the report.

Macro Parameters

The macro uses macro parameters to define the three macro variables that are used in the simulation.

Parameter	Represents
new	name of output data set
old	name of input data set
amount	amount of savings

This allows you to identify the input data set that you want to re-price, create a unique output data set name, and specify a minimum savings amount. The macro parameters are defined in parentheses in the %macro statement below:

    %macro pricing(new,old,amount);

39

You enter values for these parameters when invoking the macro, as in:

    %pricing(reprice2002,paid2002,200)

In this example you are telling SAS to output a new SAS data set called "reprice2002", read the "paid2002" paid invoice data set, and use a value of $200 to output only re-priced products where the savings are greater than or equal to $200. The macro processor resolves the macro variables as follows:

Original statements:

```
        data &new;
         set &old;

         .

         .

         .

        if savings > &amount;
```

Resolved statements:

```
        data reprice2002;
         set paid2002;

         .

         .

         .

        if savings > 200;
```

Record Selection

In situations where you want to subset the input data set to focus on a select set of records, a subsetting IF statement can be used to invoke a macro containing the specific identifiers of interest. In our simulation, we are only interested in selecting paid invoices for a set of preferred vendors. The macro shown below, which is stored in a separate file, contains the set of preferred vendors identifiers used in our simulation.

Macro VENDORLIST – ex. 1

```
%macro vendorlist;

  vendor = 40064
or vendor = 71055

%mend vendorlist;
```

Macro PRICING invokes macro VENDORLIST with the following statement:

Original statement

        if %vendorlist;

Resolved statement

```
        if  vendor = 40064
        or vendor = 71005;
```

To specify a different set of preferred vendors, or even a single vendor, simply change the vendor number values by modifying VENDORLIST, as in:

Macro VENDORLIST – ex. 2

```
%macro vendorlist;

vendor = 55804

%mend vendorlist;
```

This time when macro VENDORLIST is invoked the resolved statement becomes:

if vendor = 55804;

Table Lookup

In our simulation, we are also only interested in re-pricing a specific set of products. However, it's very likely that this set of products will change over time. Rather than hard-code them into the program, you can add flexibility by packaging the product identifiers separately in a different file. You can maintain many sets of product identifier's that vary from one analysis to another. You simply select the set you want to use when you run the analysis, without having to modify the program. The "table look-up" concept uses the macro language as the means of packaging external lists. The macro shown below, which is stored in a separate file, contains the set of product codes used in our simulation.

Macro PRODLIST

```
%macro prodlist;

  product = "A10239" or
  product = "A10251" or
  product = "A10280" or
  product = "A10297"

%mend prodlist;
```

Macro PRICING invokes macro PRODLIST with the following statement:

Original statement

if %prodlist then do;

Resolved statement

```
If product = "A10239" or
   product = "A10251" or
   product = "A10280" or
   product = "A10297" then do;
```

BY Statement Processing

In the examples illustrated above, we can see that the analysis can shift from a single vendor to multiple vendors. To support this kind of flexibility, the mechanism that produces the report output must be capable of dealing with multiple entities. The BY statement processing technique is designed to allow for the processing of an input data set regardless of whether it contains data for one or several entities, i.e., vendors. Macro PRICING contains a PROC PRINT step that is built around a BY statement. If the input data set contains observations for a single vendor the report will appear as follows:

Listing of Selected Products From: reprice2002 - Date: 27MAY03
Products Where the Simulated Difference Between
Actual Cost and Simulated Cost is Greater Than or Equal to $200

---Preferred Vendor=40064 ---

Preferred Vendor	Vendor Name	Product	Wholesale Price	Actual Cost	Simulated Cost	Simulated Savings
40064	Smith Bros.	A10239	$9,000	$10,500	$9,900	$600
40064	Smith Bros.	A10251	$10,000	$12,000	$11,000	$1,000
40064	Smith Bros.	A10280	$11,000	$13,000	$12,100	$900
40064	Smith Bros.	A10297	$12,000	$15,000	$13,200	$1,800
			----------	----------	----------	----------
vendor			$42,000	$50,500	$46,200	$4,300
			=======	=======	=======	======
			$42,000	$50,500	$46,200	$4,300

If the input data set contains observations for multiple vendors the report output will appear as follows:

Listing of Selected Products From: reprice2002 - Date: 27MAY03
Products Where the Simulated Difference Between
Actual Cost and Simulated Cost is Greater Than or Equal to $200

--- Preferred Vendor=40064 ---

Preferred Vendor	Vendor Name	Product	Wholesale Price	Actual Cost	Simulated Cost	Simulated Savings
40064	Smith Bros.	A10239	$9,000	$10,500	$9,900	$600
40064	Smith Bros.	A10251	$10,000	$12,000	$11,000	$1,000
40064	Smith Bros.	A10280	$11,000	$13,000	$12,100	$900
40064	Smith Bros.	A10297	$12,000	$15,000	$13,200	$1,800
			----------	----------	----------	----------
vendor			$42,000	$50,500	$46,200	$4,300

--- Preferred Vendor=71055 ---

Preferred Vendor	Vendor Name	Product	Wholesale Price	Actual Cost	Simulated Cost	Simulated Savings
71055	Hansen Ind.	A10239	$9,100	$10,500	$10,192	$308
71055	Hansen Ind.	A10251	$10,200	$12,200	$11,424	$776
71055	Hansen Ind.	A10280	$11,050	$13,300	$12,376	$924
71055	Hansen Ind.	A10297	$12,250	$15,500	$13,720	$1,780
			----------	----------	----------	----------
vendor			$42,600	$51,500	$47,712	$3,788
			=======	=======	=======	======
			$84,600	$102000	$93,912	$8,088

Unbundled Programs

Programmer/analysts have been known to create long, complicated programs that produce a vast quantity of output and/or require considerable processing time to complete. An example might be a program that:

1. Reads a raw data file into a SAS data set
2. Sorts the SAS data set

3. Merges with another SAS data set
4. Summarizes the data
5. Prints several reports

As you can see, this single program really consists of five discrete tasks. What happens if after reviewing the output, you conclude that a raw data field was incorrectly processed? In all likelihood you would rerun the program, thereby reprocessing the same data in steps 2-5 that were needlessly processed in the first run. Unbundled programming is an alternative coding practice that strives to simplify and disaggregate data manipulations into small, self-contained segments. This approach can produce cleaner more easily maintained code and add flexibility to processing.

The general principles of unbundled programming are to:
- Create programs that accomplish one primary task, such as reading a raw data file.
- Separate DATA and PROC steps where possible, particularly when the PROC steps are used exclusively to produce reports.

Working with Large Data Sets

Efficient processing of large files requires the programmer/analyst to exercise sound judgment when selecting among alternative approaches for writing code. Several characteristics of the SAS System should be considered when processing large files.
- SAS tends to pass the data many times.
- SAS data sets may end up larger than the original data file.
- Some applications may require frequent sorting of the data.

Recommendations for dealing with these characteristics are outlined below.

Multiple Passes of Data

In a common and simple program, SAS would typically make at least three input/output passes through the data: reading the raw data, outputting a SAS data set, and inputting the SAS data set to a procedure. The following techniques can decrease the number of passes through the data:
- Build data sets containing different subsets of the data in a single DATA step.
- Use the FIRSTOBS= and OBS= data set options to perform limited data selection in procedures.
- Use the APPEND procedure rather than a DATA step to concatenate SAS data sets when the data values in them do not require modification.

Data Expansion

Using the following techniques when building SAS data sets can minimize the problem of data expansion:
- Input only the fields necessary for the analysis.
- Reduce the number of variables with the KEEP and DROP statements or KEEP= and DROP= data set options.
- Store numbers as character variables when the numbers represent classifications (e.g., '0' = female; '1' = male).
- Use a LENGTH statement to specify the size of data fields in the SAS data set. (Caution should be used when developing the initial length specifications so that truncation problems do not occur in subsequent analyses.)

Frequent Sorting

Sorting large data sets is time-consuming and resource-intensive. In addition, processing the data set by several different subgroups may require multiple sorts. The following techniques are helpful in reducing the number of times a sort is required:

- When a BY statement is used to process a data set in subsets, SAS expects the observations to be in the same order as though the SORT procedure has been used. If the observations are already in the correct order, a PROC SORT step is not required.
- If observations with the same BY value are grouped, but the groups are not in alphabetical or numeric order, use the NOTSORTED option with the BY statement.
- Whenever possible, use procedures like SUMMARY, FREQ, and CHART, which collect statistics on subgroups of data without requiring a sort.

In addition, SAS does not sort a data set "in place". Instead, it makes a new copy of the data set. There must be adequate space to store both the new and the old data sets at least momentarily.

Planning Checklist

Careful planning before running a time consuming program can prevent costly losses due to errors. The checklist below provides several helpful suggestions for avoiding problems.

Computer Resources

- Investigate any hardware environment limitations, such as memory or file storage space that might cause a job to fail.
- Store the data in the form of a SAS data set instead of reading the raw data with each analysis.
- Save intermediate files created in a job. If the job fails in the middle, it will not have to be rerun from the beginning.
- Make test runs with small amounts of sample data before running a job against the entire file.

SAS Programming

- In a series of nested IF/THEN/ELSE statements or in a SELECT group, order the conditions so that the ones most likely to be true come first. SAS skips the remaining statements in the series or group after it finds a true condition.
- If a program contains a long series of mutually exclusive conditions, a SELECT group makes the program easier to read and debug than a series of IF/THEN/ELSE statements.
- Include code in the SAS statements to check for and reject data that could invalidate the results.
- If an expression is required in more that one statement, assign the result of the expression to a variable and use the variable in later statements.
- Assign values to constants in a RETAIN statement rather than in an assignment statement.
- Array processing in the DATA step saves programming time and makes a program more flexible; it does not save processing time.

Input/Output

- DATA '_NULL_' should be used in the DATA step when programming is required but an output data set is not required.
- When working from a permanent SAS data set, it may be possible to start the program with a PROC step, rather than setting the data set in a DATA step.

- Use conditional INPUT techniques whenever possible: read in a field with a trailing @ and check its value; if the record is desired, read in the rest of the variables.
- Place subsetting IF statements as close to the beginning of a DATA step as possible. Keep in mind that when an observation does not meet the subsetting condition, SAS returns to the beginning of the DATA step immediately and does not execute statements following the subsetting IF for that observation.
- There is no difference in column versus formatted input in terms of efficiency.
- Computing variables in a DATA step uses fewer resources than inputting them.

Report Writing

- Consider using PUT statements rather than PROC PRINT.
- Use a FORMAT statement with PROC PRINT. When formats are not specified, the PRINT procedure takes extra time to search for the best format for each variable. When formats are given, the printout is uniform from page to page.

Summary - Key Points

- That first glimpse at the data can be very revealing.
- Account for all the raw data records.
- Design the analysis data set(s) based on the questions to be answered.
- A significant amount of programming may be required to build the analysis data set(s).
- Use appropriate programming techniques to efficiently manage your code and process your data.

6. Describing the Data

Introduction

Before jumping into the analysis dataset(s) to produce the final reports, we use this step to scrutinize the data in a variety of ways. We look for extreme and unusual values, test whatever assumptions we still have, and otherwise examine the data to assess their completeness, accuracy, and validity. Only after convincing ourselves that these three attributes are accounted for are we in a position to proceed with final reporting.

Assessing data quality through a validation process is a necessary but often difficult task. Measurements of data quality are often subjective, poorly defined, or constrained by a lack of insight into how the data were originally captured. Data collection processes can impart error by their very nature. In spite of these challenges, there are several concrete steps that you can take to examine your data and thereby deduce overall quality.

Data Completeness - Part II

This is a continuation of the initial data completeness check introduced in chapter 5. We'll take a deeper look into the data in search of potential problems related to missing or incomplete values within key fields.

Frequency Distributions

For a field that should contain a limited number of values, create a frequency table that describes your data by reporting the distribution of values within the field. The "ch5ex" data set, containing the same 13 observations used in chapter 5, is input to the FREQ procedure shown below.

```
proc freq data=ch5ex;
   tables gender;
```

The following frequency table is produced showing all observed values for the gender field. In this example there are no surprises, in that only the expected values of "F" and "M" are present, and they are present for all observations. We also see that the data set contains thirteen values, which is also consistent with our expectation. Therefore, we have an initial indication that this data set does contain the expected number of observations and that each observation has a value for the gender field.

			Cumulative	Cumulative
gender	Frequency	Percent	Frequency	Percent
F	6	46.15	6	46.15
M	7	53.85	13	100.00

The SAS System

The FREQ Procedure

The SAS FREQ procedure can also easily produce *n*-way frequency tables, in addition to the one-way table displayed above.

If your data contained blank or invalid values, the frequency table would clearly indicate their presence. In this example we see one occurrence of the value "G" and an indication of one observation with a missing value for gender.

```
                          The SAS System

                         The FREQ Procedure

                                          Cumulative    Cumulative
      gender      Frequency     Percent   Frequency     Percent

      F               5          41.67         5          41.67
      G               1           8.33         6          50.00
      M               6          50.00        12         100.00

                        Frequency Missing = 1
```

If you encounter missing values you will need to assess how widespread they are and how you will deal with them. If the number of missing values for a particular field is small you may decide to exclude those observations. If the number is large, it may make the field unusable and thereby confound or jeopardize your analysis plan. Always keep in mind the potential impact that missing numeric values can have on calculations. This impact will be demonstrated in the next section.

Descriptive Statistics

For fields containing numeric data, there are several SAS procedures that will create a report of descriptive statistics. These reports are useful for identifying data anomalies such as extreme or out-of-range values. Descriptive statistics such as the mean, minimum, and maximum are easily produced.

In the first example, we will use the MEANS procedure for describing the numeric data in the ch5ex data set. In its simplest form, PROC MEANS prints the *n*-count (number of non-missing values), the mean, the standard deviation, and the minimum and maximum value of every numeric field in the data set.

```
proc means data=ch5ex;
run;
```

With the exception of the ID variable, the information below is useful. We can review the descriptive statistics for age, height, and weight to see if they conform to our expectations for reasonableness. Obviously for a variable such as ID, these statistics are not meaningful.

```
                             The SAS System

                           The MEANS Procedure

     Variable    N       Mean       Std Dev       Minimum        Maximum

     ID          13    557431521    186756712    120896346      809217374
     age         13    36.1538462   11.9711191    22.0000000     59.0000000
     height      13    68.6923077    4.3852901    60.0000000     74.0000000
     weight      13   163.076921    34.1893296   100.0000000    225.0000000
```

We can add several statements to the MEANS procedure to restrict calculations to a subset of variables, specify statistics, and calculate statistics for grouped observations. In the example below, we specify that only the mean, minimum, and maximum statistics will be calculated for the variables age, height and weight. In addition, these statistics will be grouped by gender.

```
proc means data=ch5ex mean min max;
  var age height weight;
  class gender;
run;
```

The intent of these modifications is to produce concise and relevant information that will provide ease of review and hopefully additional insight to the data. With the addition of the gender grouping, we now have a much more meaningful set of statistics to review for reasonableness.

The SAS System

The MEANS Procedure

gender	N Obs	Variable	Mean	Minimum	Maximum
F	6	age	34.1666667	22.0000000	52.0000000
		height	65.0000000	60.0000000	70.0000000
		weight	133.3333333	100.0000000	150.0000000
M	7	age	37.8571429	23.0000000	59.0000000
		height	71.8571429	70.0000000	74.0000000
		weight	188.5714286	170.0000000	225.0000000

Now let's reintroduce missing values to better understand their impact. If a single observation in the "ch5ex" data set has a missing value for weight (in this case due to invalid data), the MEANS procedure output will change accordingly, as shown below.

The SAS System

The MEANS Procedure

gender	N Obs	Variable	Mean	Minimum	Maximum
F	6	age	34.1666667	22.0000000	52.0000000
		height	65.0000000	60.0000000	70.0000000
		weight	133.3333333	100.0000000	150.0000000
M	7	age	37.8571429	23.0000000	59.0000000
		height	71.8571429	70.0000000	74.0000000
		weight	189.1666667	170.0000000	225.0000000

Only one number has changed. You'll notice that the mean weight for males has increased from 188.57 pounds to 189.16 pounds. Furthermore, the 189.16 statistic is calculated using only the six observations with non-missing values for weight rather than all seven observations for males. The observation count above for males is the number of observations where gender = 'M', not the count of observations used to calculate the mean weight.

In a second example, we'll introduce invalid and missing values for gender to the "ch5ex" data set and then rerun the MEANS procedure.

```
                              The SAS System

                            The MEANS Procedure

              N
gender       Obs    Variable        Mean       Minimum       Maximum

F             5     age        33.0000000    22.0000000    52.0000000
                    height     64.0000000    60.0000000    68.0000000
                    weight    130.0000000   100.0000000   150.0000000

G             1     age        40.0000000    40.0000000    40.0000000
                    height     70.0000000    70.0000000    70.0000000
                    weight    150.0000000   150.0000000   150.0000000

M             6     age        36.0000000    23.0000000    59.0000000
                    height     71.8333333    70.0000000    74.0000000
                    weight    189.1666667   170.0000000   225.0000000
```

Reviewing these results we observe multiple changes to output statistics and the introduction of a completely invalid gender category ("G"). We can also note that the number of observations for legitimate gender values has been reduced for both males and females due to the bad input data.

Remember that even though the impact of missing and invalid values has been somewhat minor in these simple examples, they have the potential for significant impact and for that reason must be carefully assessed <u>before</u> proceeding with calculations.

Another powerful but easy to use SAS procedure is PROC UNIVARIATE. This procedure will produce one page of statistical information for <u>each</u> numeric variable specified. Be careful what you ask for!

```
proc univariate data=ch5ex;
  var age height weight;
run;
```

Note that the UNIVARIATE procedure output is very extensive. However, you can easily glean essential information such as mean, standard deviation, variance, as well as the listing of the five lowest and highest observations in the "Extreme Observations" section of the report. Only the report output for the age variable is displayed below.

```
                              The SAS System

                          The UNIVARIATE Procedure
                               Variable: age

                                 Moments

N                        13       Sum Weights                13
Mean              36.1538462      Sum Observations          470
Std Deviation     11.9711191      Variance            143.307692
Skewness           0.5497441      Kurtosis           -0.7550031
Uncorrected SS         18712      Corrected SS       1719.69231
Coeff Variation    33.111606      Std Error Mean      3.32019105
```

```
                        Basic Statistical Measures

        Location                            Variability

Mean        36.15385        Std Deviation           11.97112
Median      35.00000        Variance               143.30769
Mode        25.00000        Range                   37.00000
                            Interquartile Range     17.00000

                      Tests for Location: Mu0=0

    Test              -Statistic-        ------p Value------

    Student's t    t      10.88909    Pr > |t|     <.0001
    Sign           M          6.5     Pr >= |M|    0.0002
    Signed Rank    S         45.5     Pr >= |S|    0.0002

                      Quantiles (Definition 5)

            Quantile          Estimate

            100% Max             59
            99%                  59
            95%                  59
            90%                  52
            75% Q3               42
            50% Median           35
            25% Q1               25
            10%                  23
            5%                   22
            1%                   22
            0% Min               22

                      Extreme Observations

    ----Lowest----              ----Highest---

    Value     Obs           Value     Obs

      22       13             40        2
      23        3             42        4
      25       11             49        1
      25        5             52        9
      27        7             59       10
```

Outlier Analysis

Outliers are usually described as "extreme" observations present in a data set, i.e., high or low valued observations that represent the tails of a distribution of values. Outliers must be considered whenever calculating statistics due to their potential to powerfully skew a distribution and thereby impact certain calculations, such as measures of central tendency (the mean in particular).

The definition of what constitutes "extreme" can be a hotly debated issue between analysts. Typically a definition is established and agreed on <u>before</u> you start examining the data. Because standard outlier definitions may not exist in your industry, you will often need to take the initiative to engage subject matter experts in a discussion of how outliers will be defined within the context of your analysis plan. For the examples in this chapter, I am using the following statement to define outliers:

> "An outlier is any observation where the value of the analysis variable is greater than two standard deviations above the mean or less than two standard deviations below the mean".

With a definition in place, we can start looking at our data.

Outlier Identification

Let's consider the SAS data set displayed below where we have values for age, height, and weight.

	The SAS System		
Obs	age	height	weight
1	49	72	310
2	99	70	150
3	23	73	180
4	42	71	190
5	25	65	115
6	32	71	175
7	27	68	145
8	35	72	195
9	52	62	140
10	59	70	225
11	25	74	170
12	39	65	150
13	22	60	100

Note that observation 1 has a value of 310 for weight and that observation 2 has a value of 99 for age. Are these outliers? At this point, we don't have enough information to know.

In addition to a tabular listing such as the report above, it's also helpful to prepare a plot of observations in order to get a different visual perspective. Plots are an effective tool for highlighting extreme observations that might otherwise be overlooked. Values for age and weight are shown in the plot below. Most values are clustered in the lower left quadrant of the plot, with the exception of one value for age = 99 and one value for weight = 310.

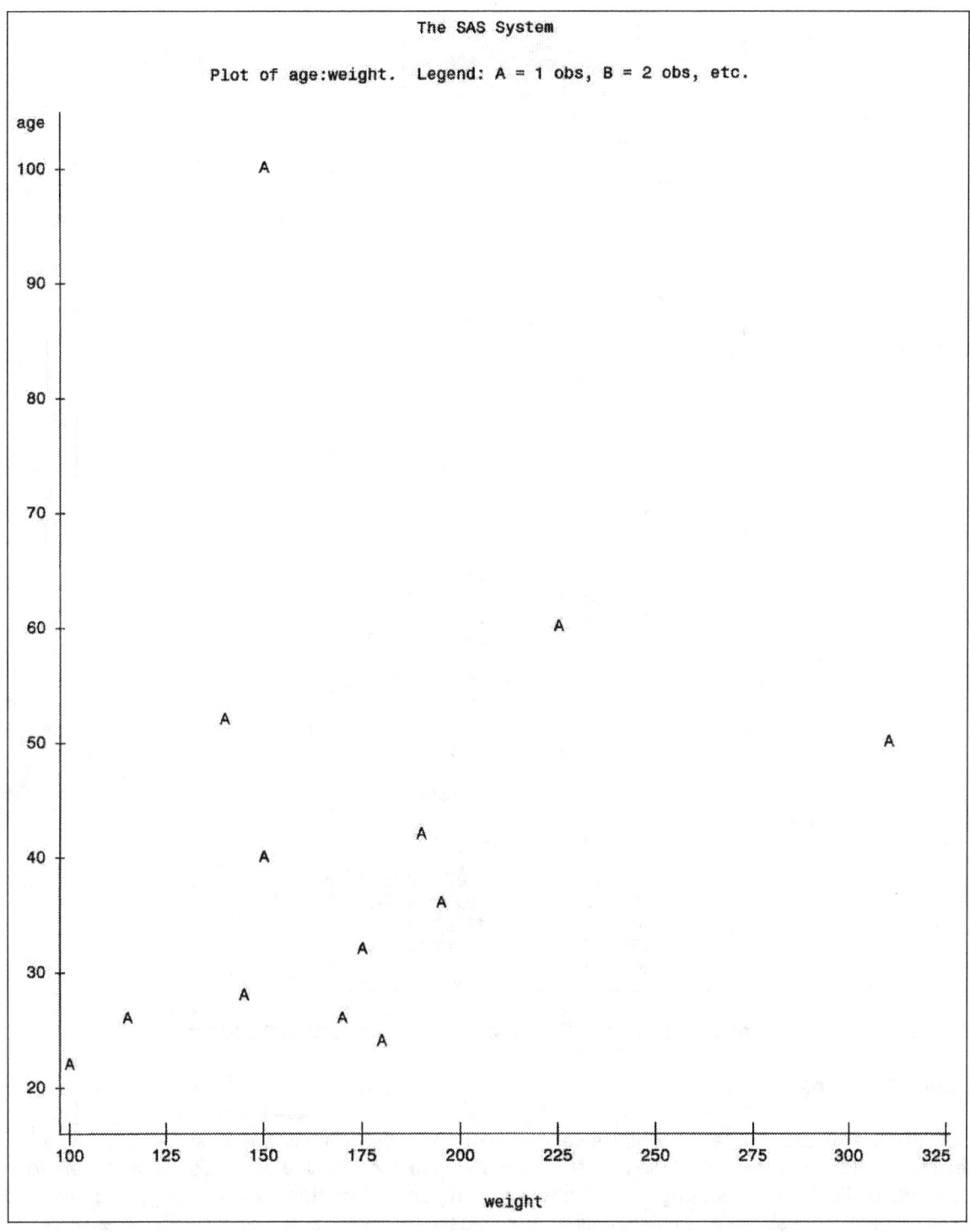

A third tool for outlier detection is the UNIVARIATE procedure that was introduced earlier in this chapter. A listing of the five highest and lowest observations is provided in the "Extreme Observations" portion of the UNIVARIATE report. Using our age, height, and weight data set as input to the UNIVARIATE procedure, the "Extreme Observations" section is displayed below for all three variables.

```
                    The UNIVARIATE Procedure
                         Variable: age

                      Extreme Observations

    ----Lowest----              ----Highest---

    Value    Obs                Value    Obs

      22      13                  42      4
      23       3                  49      1
      25      11                  52      9
      25       5                  59     10
      27       7                  99      2


                        Variable: height

                      Extreme Observations

    ----Lowest----              ----Highest---

    Value    Obs                Value    Obs

      60      13                  71      6
      62       9                  72      1
      65      12                  72      8
      65       5                  73      3
      68       7                  74     11


                        Variable: weight

                      Extreme Observations

    ----Lowest----              ----Highest---

    Value    Obs                Value    Obs

     100      13                 180      3
     115       5                 190      4
     140       9                 195      8
     145       7                 225     10
     150      12                 310      1
```

Once again, the observations with age = 99 and weight = 310 surface in these reports.

Outlier Thresholds

Now that we know our data set contains several "extreme" observations, we must decide if these really constitute outliers. Using the MEANS procedure, we can produce the statistics necessary to calculate outlier thresholds using our definition of the mean +/- two standard deviations. For the sake of thoroughness, we'll calculate outlier thresholds for all three of our analysis variables, even though it doesn't appear that any height outliers are present in the data set.

```
                    The SAS System

                   The MEANS Procedure

        Variable          Mean          Std Dev


        age           40.6923077       21.1871998
        height        68.6923077        4.3852901
        weight       172.6923077       53.1748446
```

Subtracting and adding two standard deviations from the mean of each analysis variable yields the following thresholds, which are rounded to the nearest integer.

Variable	Lower Threshold	Upper Threshold
age	-2	83
height	60 inches	77 inches
weight	66 lbs.	279 lbs.

Now we have something concrete to work with. Any observation with values for the three analysis variables that fall <u>outside</u> of lower and upper thresholds will be classified as an outlier observation. Note that it is possible for a single observation to achieve outlier status based on its value for <u>any</u> analysis variable. This means an observation can simultaneously be an age, height, and weight outlier. This possibility becomes important if we decide to eliminate outliers from the analysis (as discussed below).

A review of the data for outliers identifies the following cases:

Obs.	Analysis Variable	Value	Reason
1	weight	310	exceeded upper threshold
2	age	99	exceeded upper threshold

We also note that there are no outliers based on height, and no observations satisfy outlier definitions for more than one analysis variable. These findings are consistent with our initial "eyeballing" of the data.

Outlier Impact on Descriptive Statistics

Now that we have precisely identified the observations with outlier values, we need to decide what to do about them, if anything. We'll start by taking a look at how the presence of outliers will impact descriptive statistics. This is demonstrated by comparing statistics derived from a data set containing outliers to the same statistics derived from the data set where outliers have been excluded.

Data Set Statistics - Outliers Included

```
                              The MEANS Procedure

   Variable    N        Mean          Std Dev        Minimum         Maximum


   age        13     40.6923077     21.1871998     22.0000000      99.0000000
   height     13     68.6923077      4.3852901     60.0000000      74.0000000
   weight     13    172.6923077     53.1748446    100.0000000     310.0000000
```

We will exclude observations with outlier values using the following logic:

```
* exclude age outliers ;
if age <= 0 or age > 83 then output ageoutliers;

*exclude weight outliers ;
else
if weight < 66 or weight > 279 then output wtoutliers;

*output non-outliers ;
else
output non-outliers;
```

Although there are many different techniques for eliminating outliers from a data set, the approach shown above has the following features:

- Each observation is output to one of three possible data sets: ageoutliers, wtoutliers, or non-outliers. The sum of observations in the three output data sets will equal the number of observations processed. This allows us to account for how each observation was categorized and to ensure that a single observation is not counted more than once.
- Each observation is first evaluated to determine if it is an age outlier. If so it is immediately written to the "ageoutliers" data set and the next observation is processed. Note that the lower age threshold is set at zero, since the theoretical lower threshold of -2 does not make sense.
- The decision to screen for age outliers before weight outliers was arbitrary in this example. It could have been done in the opposite manner. The elimination order becomes an issue when a single observation satisfies more than one outlier definition.
- If an observation is not an age outlier, it will be evaluated to determine if it is a weight outlier. If so it is immediately written to the "wtoutliers" data set and the next observation is processed.
- All other observations are by definition non-outliers and are written to the "non-outliers" data set.

Data Set Statistics - Outliers Excluded

		The MEANS Procedure			
Variable	N	Mean	Std Dev	Minimum	Maximum
age	11	34.6363636	12.3066870	22.0000000	59.0000000
height	11	68.2727273	4.6495357	60.0000000	74.0000000
weight	11	162.2727273	36.5625243	100.0000000	225.0000000

It's easy to see that the result of excluding high outliers is a dramatic reduction in mean age and weight. Removal of extreme observations also reduces the amount of variability in the data set which if reflected in a smaller standard deviation. Note also that the mean and standard deviation for height changed because they were calculated on a different data set (11 observations versus 13).

Data Validity

Assessing data validity is more difficult than measuring completeness. In this case our challenge is more subjective and therefore a weaker test. We are interested in evaluating whether a measure appears valid "on its face", i.e., face validity. How do we determine if the observed values are legitimate for the context in which they appear? Data values that clearly exceed the range of possible values are certainly invalid, such as a value of 188 for age. In other cases, it's not quite so obvious. For example, we may observe a value of 55 for age. This value has passed the completeness test because it is present and readable. But is it valid? To answer that question we need to view the value in context. If our data set represents a population of NFL football players, then we could easily conclude that a value of 55 for age is invalid. If our data set represents a population of early retirees, we might draw a different conclusion. Still in other cases we observe values that initially appear legitimate, but when further scrutinized, fail the validity test. For example, if we were reviewing a frequency distribution of student grades where the possible values include "A", "B", "C", "D", and "F", we would not expect to see a value of "V". A value of "V" in this context has no relevant meaning.

As a matter of process, begin by preparing and reviewing frequency tables to compare all observed values against expected values (e.g., from the data dictionary). The distribution frequency should be assessed in terms of known information about the population. For example, in the first frequency distribution above we observe approximately half females and half males. Is this what you would expect to see? If the population were known to be heavily skewed in the direction of one gender (e.g., law enforcement or nursing), then these results would be suspect. Also, as previously notes, we observe one occurrence of the value "G" for gender in the second frequency distribution. Unless you know that "G" is a legitimate value, this finding would be evidence of a data quality issue.

Data Accuracy

Now we're faced with the most difficult test. Are the data accurate? This is a measure of content validity that requires us to understand the process by which data are defined and assembled. In other words, were the data captured correctly? For example, if the file contains an observation with a value or 43 for age, how do you know that 43 is the correct age? What if the real age is 34 and the digits were reversed during data entry? You may never know. In practice we may have limited or no information about how the data were originally captured.

Efforts to assess accuracy depend on the availability of up-to-date documentation about how fields are specified and the production processes that populate those fields. Without adequate documentation we can easily misinterpret fields based on their names. For example, how is the age field specified? It is age at date of capture, or age since a point in time? In the latter case, the value must be calculated using date of birth and the comparison date. What do we know about the process by which those values are captured? Was there any type of data entry validation? Are there events that trigger filling in a field, and are those events mandatory or optional?

Summary - Key Points

- Understand your data before attempting to produce final reports from it.
- Look at your data from multiple perspectives, e.g., "a graph is worth a thousand numbers".
- Dig for outliers.
- If you find outliers, figure out how they will impact your analysis.
- Be prepared to respond to inquiries about the completeness, accuracy, and validity of your data. Someone may ask.

58

7. Answering the Questions

Introduction

Even though we have a solid foundation for our analysis by this stage of the project, we may still need to build upon that foundation through additional data manipulation. This can take the form of creating additional variables, which might require the use of complex logic, transforming the data structure, or otherwise summarizing or reformatting the data for reporting purposes. Simply put, this part of the process is where we turn data into information so that the analysis questions can be answered.

Manipulating the Data

Bits of data are like LEGO® blocks. They can be assembled into an amazing number of different structures and shapes. However, that doesn't mean that the resulting shape meets the design specifications. If the requirement was to build a LEGO <u>car</u> but you ended up with a LEGO <u>star</u>, you haven't accomplished your objective. Data manipulation is a lot like assembling individual LEGO blocks (i.e., data) into a shape or structure that will satisfy the design requirements (i.e., questions to be answered). You won't be able to answer the questions if the data are not structured appropriately. There are many data manipulation tools and techniques available. Several considerations to keep in mind when working with your data are offered below.

Operators

Because operators are such a fundamental part of data manipulation, a brief introduction is offered here. Operators are used to construct formulas used in calculations, and for logical and comparative expressions. They are easy to use, and easy to use incorrectly. In some cases there are multiple words or symbols for a single operator. In the table below I have chosen the format that I believe offers the clearest indication of your intent.

Description	Operator	Example
Numeric Operators:		
Plus sign	+	A + B
Minus sign	-	A - B
Multiplication	*	A * B
Division	/	A/B
Exponentiation	**	A**B
Logical Operators:		
And	AND	A AND B
Or	OR	A OR B
Not	NOT	NOT A
Comparison Operators:		
Equal to	=	A = B
Not equal to	NE	A NE B
Greater than	>	A > B
Less than	<	A < B
Greater than or equal to	GE	A GE B
Less than or equal to	LE	A LE B
Equal to one of a list	IN	A IN (1,2,3)

Many of these operators will be used in the examples that follow.

Writing Formulas and Logical Expressions

Many formulas and logical expressions can be written in multiple ways, all equivalent. When writing these statements strive to keep them simple and intuitive, rather than complicated and confusing. Not only will this aid in testing and documentation, it will also help ensure that the logic you intend is the logic that actually gets coded.

Use of Parentheses

Consider the following example. Both formulas calculate a value for "cost". Each formula produces exactly the same result even though they are structured differently.

cost=hours*rate*1000/2080*1.32;

cost = hours * (((rate * 1000)/2080) * 1.32);

Translating the first formula into words would produce "cost is equal to hours multiplied by rate multiplied by 1000 divided by 2080 multiplied by 1.32". This description is accurate but doesn't tell us much about what this formula really represents. The second formula provides several clues. When translated into words it suggests that "cost is equal to the product of the annualized rate and overhead loading factor multiplied by the number of project hours". Perhaps that interpretation is a bit of a stretch, but I think the concept is clear. The second formula certainly imparts more structure and organization, and is therefore more interpretable, than the first.

Tip: When adding parentheses to a formula, remember that they must be balanced, i.e., equal numbers of "(" and ")".

Even though parentheses can add clarity and organization to your formulas, they also have a dark side. Simply keeping them balanced does not guarantee that they might not introduce unintended error. Look the four formulas below.

result = 5**5*2/5;

result = (5**5)*(2/5);

result = ((5**5)*2)/5;

result = 5**(5*2/5);

All of these formulas will execute without error. However, one of them will return a result different from the others. If your intent is to "multiply 5 raised to the 5th power by two-fifths" then any of the first three formulas will correctly calculate a value of 1,250, with the second formula providing the clearest indication of what you want to calculate. The last formula returns a result of 25, because the use of parentheses has instructed the computer to "raise 5 to the power of 2".

Tip: Translate your formulas into words to see if they still make sense.

In the next example we want to calculate the percentage increase for someone who received a raise of $6000 to his or her $60000 salary.

percent = ((newsalary - oldsalary)/oldsalary)*100;

percent = (newsalary - oldsalary/oldsalary)*100;

percent = newsalary - oldsalary/oldsalary*100;

The first formula correctly calculates a percentage increase of 10%, based on "newsalary minus oldsalary divided by oldsalary multiplied by 100". The second formula calculates a value of 6599900 based on "newsalary minus 1 multiplied by 100". Finally, the third formula calculates a value of 65900 based on "newsalary minus 100".

```
percent = ((newsalary - oldsalary)/oldsalary)*100;
        = ((66000 - 60000)/60000)*100
        = (6000/60000)*100
        = 10%
```

```
percent = (newsalary - oldsalary/oldsalary)*100;
        = (66000 - 60000/60000)*100
        = (66000 - 1)*100
        = 6599900

percent = newsalary - oldsalary/oldsalary*100;
        = 66000 - 60000/60000*100
        = 66000 - (1*100)
        = 66000 - 100
        = 65900
```

Tip: Remember that the computer will do exactly what you tell it to.

Division

Special care should be taken when formulas contain division. If you inadvertently divide by a missing value or zero, the operation will fail and the SAS log will contain one of the following notes.

> NOTE: Missing values were generated as a result of performing an operation on missing values.
>
> NOTE: Division by zero detected at line 935 column 24.

It is a good practice to test the denominator before dividing in order to avoid these problems. There are several methods for doing this. The most intuitive approach is to use an expression with operators, such as the example below.

```
if denominator > 0 then ratio = numerator/denominator;
else ratio = .;
```

Tip: A more efficient approach is to use the denominator variable directly as the condition because if it is 0 or missing, it will represent a false logical value in its own right.

```
if denominator then ratio = numerator/denominator;
else ratio = .;
```

If/Then/Else Logic

Another potential quagmire is the commonly used, although somewhat dated, If/Then/Else statements. A number of issues must be considered when using this approach. In the first example, we have a simple comparison that uses the value of "type" to assign a value to "newcolor".

```
if type = 'A' then newcolor = 'Orange';
else
if type = 'B' then newcolor = 'Purple';
```

This works fine, as long as you encounter only values of 'A' or 'B' for the type variable. What happens if the value of type is 'C'? The answer is: "The value of newcolor will be blank" because the condition (type = 'C') does not occur in the logic statements. Fix this limitation by adding the "else" component to the statements.

```
if type = 'A' then newcolor = 'Orange';
else
if type = 'B' then newcolor = 'Purple';
else
newcolor = 'Other';
```

Or, what happens if the value of type is lower case 'a'? The answer is the same as above, the value of newcolor will be blank because the condition (type = 'a') does not occur in the logic statements.

A third situation to avoid is related to the process by which the length of character variables is established. In the following example, the value of the "type" is used to assign a value to "newcolor".

```
If type = 'A' then newcolor = 'Red';
else
if type = 'B' then newcolor = 'Blue';
else
newcolor = 'Other';
```

If the first observation processed has a value of 'A' for type, then the value of newcolor will be set to 'Red' and the <u>length</u> of newcolor will be set to **3**. Likewise, if the first observation processed has a value of 'B' for type, then the value of newcolor will be set to 'Blue' and the length of newcolor will be set to **4**. You can probably see where this is going. What happens if we encounter type = 'A' in the first observation following by a value of 'B' in the second observation? As we already know, the value of newcolor for the first observation will be 'Red', but for the second observation, the value for newcolor will be truncated to 'Blu', due to the limit set by the LENGTH attribute. This problem can be solved by expanding the length of 'Red' and 'Blue' to five characters, which is the length of the largest value ('Other') that can be assigned to newcolor.

```
if type = 'A' then newcolor = 'Red ';
else
if type = 'B' then newcolor = 'Blue ';
else
newcolor = 'Other';
```

Tip: Use a LENGTH statement <u>prior</u> to the if/then/else statements to initially establish an adequate length for newcolor.

```
length newcolor $ 5;
if type = 'A' then newcolor = 'Red';
else
if type = 'B' then newcolor = 'Blue';
else
newcolor = 'Other';
```

Comparisons

When writing logical comparisons, take the time to think carefully about the statements you code. First impressions can be deceptive. In the following examples we're comparing dates to various start and end points that define a time period. The first expression tests whether the value for "date" falls into the time period defined by the value of "start" and "end".

```
if (start  <= date <=  end) then do;
  other statements
end;
```

This expression reads as "If the value for date is greater than or equal to the value for start <u>and</u> the value for date is less than or equal to the value for end then do something". Using the sample data displayed below, we can see how various values for "date" affect the comparison and whether the condition for further processing is met.

start	date	end	condition met?
July 1, 2003	July 15, 2003	July 31, 2003	yes
July 1, 2003	July 15, 2002	July 31, 2003	no
July 1, 2003	July 31, 2003	July 31, 2003	yes
July 1, 2003	Aug. 1, 2003	July 31, 2003	no

This expression is easy to read, is interpretable, and functions as intended.

How do we change the expression to identify dates that fall <u>outside</u> of a time period? Shouldn't it be as simple as reversing the comparison operators? If so, then the following expression should work.

```
if (start > date > end) then do;
    other statements
end;
```

The intent is to create an expression that reads "If the value for date is less than start (which puts it before the time period) or the value for date is greater than end (which puts it after the time period) then do something". This expression fails because a single value cannot be simultaneously less than <u>and</u> greater than two different numbers. Plugging in some test data demonstrates the problem.

start	date	end	condition met?
July 1, 2003	June 15, 2003	July 31, 2003	no
July 1, 2003	June 30, 2002	July 31, 2003	no
July 1, 2003	Aug. 1, 2003	July 31, 2003	no
July 1, 2003	Aug. 15, 2003	July 31, 2003	no

What we really need is an OR condition such as:

```
if (date < start or date > end) then do;
    other statements
end;
```

This expression will give us what we want as the sample data shows.

start	date	end	condition met?
July 1, 2003	June 15, 2003	July 31, 2003	yes
July 1, 2003	June 30, 2002	July 31, 2003	yes
July 1, 2003	Aug. 1, 2003	July 31, 2003	yes
July 1, 2003	Aug. 15, 2003	July 31, 2003	yes

Tip: If the logic is complex, you should create test data to run against the code in order to prove that the logic does what you think it does.

Testing

You may be wondering why the topic of testing hasn't appeared until chapter 7. Shouldn't we be testing at every phase of the process? The answer, of course, is a resounding "yes". Testing is not limited to the data manipulation tasks discussed in this chapter. Testing should be an inherent and deliberate part of each program you create or modify, where the objective is to verify that your program does what you think it does.

Since entire books are dedicated to testing I will simply offer a few suggestions for this important and essential part of the analytic work process.

- Use the test folder structure discussed in chapter 3. Have separate folders for test data and test code.

- Develop and maintain test data. Don't rely on extracting a few records from the raw data file to use as test data. It's unlikely that a few records from a production file will contain enough conditions to thoroughly test your code.
- Test for a variety of conditions:
  - o Out of range values, for both numeric and character fields
  - o Negative values
  - o Overflow conditions
  - o Division and multiplication by zero
  - o If your reports span multiple time periods, such as month-end, quarter-end, and year-end, then you must check them all.
- Developing good test data can take a lot of time and attention to detail. Keep it for use when modifying the program in the future.
- Start small and build up by testing sections or modules before entire programs. The same concept applies to testing complex calculations.
- Save tested programs before making major modifications. It gives you a place to return to if elaborate changes get out of hand.
- If you strive to keep your programming simple and well documented, then testing will be easier.
- If you use exotic code with unnecessary complexity to impress others, testing will take longer than it should or it will be incomplete, which is worse.
- Save test results as part of project documentation.
- Get others to participate by having code reviews. Don't consider a code review to be successful unless you have uncovered errors, inefficiencies, or an alternate way of doing something.

As a final comment, I have found that those who can truly put their ego aside while testing, are more likely to garner the benefits that rigorous and comprehensive testing can provide. If your testing efforts aren't surfacing errors, you probably aren't testing hard enough.

Tip: Testing is as much art as science. Approach it with creativity.

Creating New Variables

The need to create new variables to support your analysis plan is a common occurrence. New variables can be created using simple expressions such as:

newtaxrate = oldtaxrate;

salestax = price * taxrate;

Functions are another useful tool for variable creation. For example the SUBSTR function can be used to extract a portion of a character value and assign it to a new variable. If you had a data set with values for FirstName and you wanted create a new variable called FirstInitial, the following statement will extract the first letter of the first name values and assign these values to the new variable FirstInitial.

```
FirstInitial = substr(FirstName,1,1);
```

Numeric functions can also be used for variable creation. The SUM function will total a set of values from a single observation.

```
total = sum(var1,var2,var3);
```

The use of functions will be discussed in greater detail below.

Character to numeric conversion is another form of variable creation. There may be situations where a value needed in a calculation is stored as a character variable. The preferred approach is to explicitly convert the character values to numeric before using them in the calculation. This is done using the INPUT function. In the example below, the character values of the variable 'taxrate' are converted to numeric values and assigned to the new variable 't_rate', which is then used to calculate 'salestax'.

```
t_rate   = input(taxrate,comma9.);

salestax = price*t_rate;
```

Finally, since we've just looked at character to numeric conversion, we should also review numeric to character conversion. The concept is similar, but the PUT function is used for explicit numeric to character conversion. Suppose you had a character variable for cost center and a numeric variable for account that you want to concatenate, i.e., create a new variable called 'LedgerLine'. Since the concatenation operator requires character values, you first convert the value of 'account' using the PUT function and assign it to 'acct' and then create the new variable 'LedgerLine'.

```
acct = put(account,4.);

LedgerLine = costcenter||acct;
```

Tip: Don't overwrite variables. For example, the following statement would change the value of the state variable from 'California' to 'CA' , resulting in a permanent loss of data.

```
if state = 'California' then state = CA';
```

This concept is even more important when grouping, as in the next section.

Grouping

Creating groups is another form of variable creation. A typical situation you might encounter is when you have a variable such as age that you want to aggregate into groups. There are many ways to accomplish this task and several guidelines to consider.

- Understand the format of the data being aggregated. For example, if you are working with a dollar field, is the value represented in whole dollars (e.g., 25) or dollars and cents (e.g., 24.99)?
- Be careful to avoid defining overlapping groups.
- Don't overwrite fields. It is better to create a new variable rather than assigning new values to an existing variable. This maintains the integrity of the existing variable in case you need to use it again in the future.
- Always create an 'Other' bucket to account for unexpected values.

In the examples that follow we'll use the sample data set shown below to create age groups.

The SAS System			
Obs	age	height	weight
1	49	72	310
2	99	70	150
3	21	73	180
4	42	71	190
5	20	65	115

6	32	71	175
7	27	68	145
8	35	72	195
9	66	62	140
10	59	70	225
11	25	74	170
12		65	50
13	4	60	100

One of the best known techniques for creating groups is through use of the if/then/else statements. Although not the most technically advanced approach, it is simple to use and understand. If we want to stratify the data set info four age groups the following statements could be used in a DATA step.

```
length agegroup $11;
if (0 < age < 6) then agegroup = 'Pre-School';
else
if (6 le age < 22) then agegroup = 'School Aged';
else
if (22 le age < 65) then agegroup = 'Adult';
else
if (65 le age < 99) then agegroup = 'Senior';
else agegroup = 'Other';
```

Printing the data set produces the following listing where a value for agegroup has been assigned for each observation.

		Age Stratified by Age Group		
Obs	age	height	weight	age group
1	49	72	310	Adult
2	99	70	150	Other
3	21	73	180	School Aged
4	42	71	190	Adult
5	20	65	115	School Aged
6	32	71	175	Adult
7	27	68	145	Adult
8	35	72	195	Adult
9	66	62	140	Senior
10	59	70	225	Adult
11	25	74	170	Adult
12	.	65	50	Other
13	4	60	100	Pre-School

Formatting

If we didn't need to reprocess each observation through a DATA step, a different approach using a format can be considered. In this case an age group format is created and then applied to output from the FREQ procedure. Instead of the comparison operators used in the if/then/else logic above, the FORMAT procedure uses age ranges in the value statement.

```
proc format;
  value agegrp  0-5     = 'Pre-School'
                6-21    = 'School Aged'
                22-64   = 'Adult'
                65-98   = 'Senior'
                other   = 'Other';
run;

title 'Formatted Age Groups';
proc freq data=agegroups;
  tables age / missing;
  format age agegrp.;
run;
```

Note that the "missing" option is used in the FREQ procedure so that frequency counts will include any missing values.

age	Frequency	Percent	Cumulative Frequency	Cumulative Percent
		Formatted Age Groups		
		The FREQ Procedure		
Other	2	15.38	2	15.38
Pre-School	1	7.69	3	23.08
School Aged	2	15.38	5	38.46
Adult	7	53.85	12	92.31
Senior	1	7.69	1	100.00

This is only one of many uses of the FORMAT procedure.

Subsetting

Another common form of data manipulation is the process of subsetting, which was introduced in chapter 5. Subsetting can be applied within a DATA step or as part of a PROC to select a specific group of observations for processing.

- Use the subsetting IF statement or a WHERE statement in a DATA step to process observations that meet the specified condition.

```
* subset observations where age is greater than 50 ;
data subset;
   set one;
   if age > 50;
run;

=======================================================

data subset;
   set one;
   where age > 50;
run;
```

- The WHERE= DATA step option is another method to limit processing to the observations that meet the specified condition.

```
* subset observations where age is greater than 50 ;
data subset;
   set one (where=(age > 50));
run;
```

- The WHERE = option and WHERE statement can also be used to limit processing in procedures.

```
title 'Age Greater Than 50';
proc print data=one;
   where age > 50;
run;

=======================================================

proc print data=one (where=(age > 50));
run;
```

- The WHERE= DATA step option can also be used to keep only the specified observations in the output data set. Use this approach if DATA step processing of all observations is required, even though only a subset will be output.

```
* output observations where age is greater than 50 ;
data subset (where=(age > 50));
  set one;
  other statements
run;
```

- Another use of the FORMAT procedure is to select specific observations when they match values in a user-defined format. In this example, only observations associated with the 'Adult' age group are processed by the PRINT procedure.

```
proc format;
  value agegrp  0-5      = 'Pre-School'
                6-21     = 'School Aged'
                22-64    = 'Adult'
                65-98    = 'Senior'
                other    = 'Other';
run;

proc print data=ch7bex (where=(put(age,agegrp.)='Adult'));
run;
```

- This technique is very useful when managing large lists of data values. If modifications to the list are required, such as adding or deleting values, only the FORMAT value statement is changed. All other downstream code (such as the PROC PRINT above) is unaffected. This is a good example of "data driven" programming.

Using Functions

This is another very broad topic that can only be briefly introduced here. Functions are a powerful tool for data manipulation, and there are many of them available. Think of functions as pre-written code that offers you a programming shortcut. Using functions in the SAS System you can:

- calculate sample statistics
- create SAS date values
- convert U.S. ZIP codes to state postal codes
- round values
- generate random numbers
- extract a portion of a character value
- convert data from one data type to another
- etc.

A SAS function can be used anywhere that you would use a SAS expression, as long as the function is part of a SAS statement. To use a function, specify the function name followed by the (one or more) function arguments, which are enclosed in parentheses.

Several examples are provided below to give you a flavor of the utility and ease of use afforded by functions.

- Use the MEAN function to calculate the arithmetic average of values. The observation in this example has values for score1, score2, and score3 as shown below. The MEAN function will calculate the average of those values and assign it to the 'avgscore' variable, which is referred to as the target variable. In this case, the value of 'avgscore' will be 3.3333333333.

68

```
SAS data:
  score1 = 1;
  score2 = 3;
  score3 = 6;

avgscore = mean(score1,score2,score3);
```

- Use the ROUND function to round to the nearest unit specified by the argument. In this example the unrounded value for 'salestax' would be 1.1712, while the value rounded to the nearest .01 equals 1.17.

```
SAS data:
   price = 19.52;

salestax = round(price*.06, .01);
```

- Use the RANUNI function to generate a random number from a uniform distribution between 0 and 1, where the function argument is the specified seed value. Executing this function three times produced the (random) results shown below.

```
x = ranuni(0);
==============

x = 0.389334351
x = 0.1542787916
x = 0.6335089261
```

- Use the SCAN function to separate a character value into words and the SUBSTR function to extract a portion of a character value. In this example, the SCAN function searches the value of 'name' for the second word that is delimited by either a blank or a comma. It returns a value of 'ROBERT' and assigns it to the variable 'firstname'. The SUBSTR function extracts from 'firstname' a one character value beginning at the first position. It returns a value of 'R' and assigns it to the variable 'firstinitial'.

```
SAS data:
   Name = 'ANDERSON, ROBERT D';

firstname = scan(name,2,' ,');

firstinitial = substr(firstname,1,1);
```

Take full advantage of functions to streamline your code and save programming time. Refer to the "SAS Language Reference: Dictionary" for more information about many other SAS functions.

Working with Dates

Date manipulation and calculations using date values can be tricky. Fortunately the SAS System provides many formats and functions to enable these tasks. SAS also stores a date value as a number, which makes calculating the number of days between two SAS dates very simple. SAS software stores dates as the number of days from January 1, 1960, to a given date. For example:

```
January 1, 1959        January 1, 1960        January 1, 1961
    -365                      0                     366
```

Because date values are stored as numbers, you can use SAS date values in numeric computations as you use any other numeric values, to include sorting based on date.

In the example that follows we'll see several SAS functions put to use to calculate age, based on the difference between birthdate and the end of the calendar year.

```
* calculate birthdate ;
birthdate = mdy(9,10,1953);

* calculate age ;
age = round((mdy(12,31,2003)-birthdate)/365.25,1);
```

This code contains several steps:

- The date of birth is calculated using the MDY function, which takes a set of numbers corresponding to the month, day, and year of an event and calculates a SAS date value. In this example, the value for 'birthdate' is -2304.
- With the end of the calendar year specified as Dec. 31, 2002, the calculation (mdy(12,31,2003)-birthdate) is equal to 12/31/2003 - 9/10/1053, or 18374. When divided by the average number of days in a year (365.25), the result is 50.305270363.
- The ROUND function rounds the value to the nearest integer, which is 50.

SAS date functions can be used anywhere within a DATA step and can also be useful in conjunction with the WHERE = clause in SAS procedures.

Reporting

The expectation at this point in the process is that the analysis dataset has been appropriately constructed to meet the needs of the analysis plan. Simply put, the data must be available and in a format to provide answers to the questions posed by the requester. Recall from chapter 2 that early in the process we have a method for defining these questions and capturing information from the requester about how the answers should be presented. We should have a good understanding of the type, format, number, frequency, and output medium that the requester has specified. The following examples illustrate the vast array of reporting options that exist.

Report Type	Format	Number	Frequency	Output Medium
Control	Table	1	Monthly	Online
Documentation	Listing	1	Per project	Paper
Draft Analysis	Chart	10	As required	Paper
Final Analysis	Chart	n/a	Quarterly	Intranet
Final Analysis	ASCII	1	Quarterly	Data file

The level of development effort between report types varies considerably due to differences in style and function. For example, when reports contain graphics, consideration must be given to the hardware necessary to support special requirements such as color output or graphics delivered to an end-user workstation. Another significant consideration is the level of presentation quality that is expected. "Fancy" reports and graphs usually require much more effort to program than simple tables or listings.

Help the readers of your report output by including informative titles and footnotes; interpretable column and row labels; appropriate dates, such as date of production or time period that the report represents; report version numbers; pagination, etc.

As you're developing the report, ask yourself these questions:

- If there is more than one report, is there a consistent look and feel across reports?
- Can I reuse this report after the next data update?
- Have I avoided hard coding dynamic information by using data driven techniques (as discussed in chapter 5)?
- Have I included too much information on one page?

- Finally, and most importantly, does this report actually answer the original question(s) that were posed?

Potential Pitfalls

There are many potential pitfalls during this stage of the project that we need to be aware of. It's also quite likely that as we near the end of the process we face a tight deadline and even the fatigue associated with a long and difficult assignment. Adhering to a defined process is one method to help mitigate these risks. You might also find that when initial answers begin to surface and are presented to your client, the questions themselves actually change. That's why the process model in chapter 1 contains a loop back to previous steps to allow for redefinition of the project objectives. Experienced analysts are not surprised when these almost inevitable iterations occur. In fact, they will anticipate this and structure programs and data to accommodate change.

Some of the common mistakes that have been illustrated in this chapter include:

- Misplaced parentheses that cause an erroneous calculation.
- Unintended division by zero.
- Faulty if/then/else logic.
- Illogical comparisons.
- Testing that lacks thoroughness or rigor.
- Failing to properly account for missing values when computing statistics.
- Data truncation.

Peeling the Onion

Have you ever heard the phrase "Data analysis is like peeling an onion."? Whoever originated that quote was describing the process of starting with the aggregate whole, and stripping away layer after layer until the core had been exposed. Each layer represents one dimension and masks the deeper layers, preventing them from being seen. As each layer is removed, a new dimension can be viewed, bringing us one step closer to the core. Can an onion conceal something under multiple layers that can't be observed from the surface? Certainly. An onion that appears fresh on the outside can be rotten at the core. The initial assessment of the whole, based on only one dimension of information, can misrepresent the true value of the onion.

When you start down the analysis path you don't know what you will find. It's like digging for gold. You never know that the next shovel full will reveal. Does each turn of the shovel produce the same result, or does one unearth a telltale sign of gold? In the world of data analysis, these nuggets come in the form of insights to the data that we use to produce information from which decisions will be made. Our job is to dig deep and observe closely so that the data will reveal itself as accurately as possible. A simple example follows to illustrate this concept.

```
|======================================================================|
```

Objective: You have been asked to analyze the data underlying a report that indicates revenue increased by 24 percent from 2001 to 2002.

Raw Data: You have access to a file containing revenue producing transaction data for both years.

Layer #1: The report shows total annual revenue as:

- 2001 $521,887
- 2002 $647,490

Layer #2: In the next level of analysis, you summarize revenue by month for each year, with the following results.

71

Month	Y2001	Y2002
1	$50,122	$173,102
2	$40,211	$ 44,030
3	$40,112	$ 44,211
4	$40,121	$ 44,220
5	$50,212	$ 55,031
6	$40,121	$ 44,201
7	$40,211	$ 44,030
8	$50,122	$ 55,211
9	$40,211	$ 44,221
10	$50,122	$ 55,030
11	$40,211	$ 33,202
12	$40,111	$ 11,001

Insights: You conclude that further exploration is required due to the following observations:

- The difference between January's is significant.
- November 2002 revenue appears low in comparison to 2001.
- December 2002 revenue appears very low in comparison to 2001.
- The other months seem to support an annual revenue increase in the neighborhood of 10 percent.

Layer #3: In order to get a better understanding of the underlying data you calculate a set of descriptive statistics by week:

Year	N	Mean	Std. Dev.	Minimum	Maximum
2001	52	$10,036	$ 44.99	$10,000	$ 10,100
2002	48	$13,460	$14,910.68	$ 1,100	$110,000

Insights: These results are rather dramatic and certainly indicate that further exploration is warranted.

- 2002 has only 48 non-missing values and should have 52 (1 per week).
- Variation within 2002 is very high.
- The range of 2002 weekly revenue is significant.

Layer #3 (next iteration): At the same level of analysis, you list weekly revenue for weeks where 2002 values are extreme.

Month	Week	Y2001	Y2002
1	1	$10,001	$ 40,000
1	2	$10,010	$110,000
1	3	$10,100	$ 1,101
11	48	$10,100	
12	49	$10,001	
12	50	$10,010	
12	51	$10,100	

Insights:

- The first week of January 2002 is approximately four times the same week in 2001. It may include a significant portion of December 2001 revenue.
- Weeks 2 and 3 appear to be an outliers.
- Data for the last week in November and the first three weeks in December 2002 is missing, understating annual results.

Layer #3 (final iteration): Descriptive statistics are recalculated after excluding extreme observations, resulting in mean 2002 revenue of $11,031.

Insights:

- Week 2 at $110,000 is approximately ten times the non-outlier mean, suggesting that it contains an extra digit (i.e., bad data).
- Week 3 at $1,100 is approximately one-tenth the non-outlier mean, suggesting that lost a digit (i.e., bad data).

Conclusions: 2002 is plagued with bad or missing data. A comparison of the original and corrected values is shown below.

Month	Week	Y2002 Original	Y2002 Corrected
1	1	$ 40,000	$11,200
1	2	$110,000	$11,000
1	3	$ 1,101	$11,010
11	48		$11,020
12	49		$11,200
12	50		$11,001
12	51		$11,000

The Truth: Using corrected data, the report was rerun and produced the following results:

- 2001 $521,887
- 2002 $573,389

The year-to-year revenue increase was actually 10 percent, not the 24 percent initially reported.

|==|

There is a lot of satisfaction in shedding light on the "obvious" to uncover a hidden truth. I suppose it's a form of detective work. When you can find the proverbial "needle in the haystack" that has eluded others, and before someone sits on it, a sense of accomplishment follows.

There's one more thing that peeling an onion and data analysis have in common. Sometimes the process will bring tears to your eyes!

Summary - Key Points

- Code can execute error free, only to manipulate the data in a different way than you intended.
- When you've finished testing, test again.
- The bottom line when reporting is to ensure that you've actually answered the question.
- Data analysis can be difficult and frustrating. Don't worry if it occasionally brings tears to your eyes.

8. Staying Organized

Introduction

Given the iterative nature of data analysis, staying organized can be a challenge. "Staying organized" is not a discrete step in the process and should not be confused with the tasks necessary to produce final project documentation. The need to maintain a certain level of control is faced at each stage of the process; it's an ongoing activity that is integral to building and maintaining the audit trail that links all elements of a project.

Building the Audit Trail

An audit trail is written documentation that traces a project from start to finish. It also allows you to distinguish among the various iterations of output that commonly occur due to the nature of the analytic process. Building and maintaining the audit trail is a fundamental part of the documentation process. In some industries it's a vigorously enforced requirement mandated by regulatory agencies. As you consider what's appropriate for your situation, also take into account the size and complexity of the projects in your organization. The documentation requirements of a small simple project are very different from what is appropriate for a large complex project.

An audit trail is an invaluable asset when you are faced with the need to validate previously achieved results. This process typically begins by recreating the sequence of events that generated the report or file in question. The audit trail should lead from the report (or file), to the job, to the program, to the programmer, to the input data set. Specific points in time should be connected to the creation of all outputs.

A number of tools that aid in the development and maintenance of an audit trail have been presented in this book. Report titles can clearly identify the project and data/program sources, the creation date, and other pertinent information. The program name identifies the project/task and the programmer. Finally, elements of the data set name include references to the programmer, report, and the source program. These naming conventions, when adhered to, provide multiple safeguards against data mismanagement. The practice of writing modular programs and saving interim results also contribute to a clear and robust audit trail.

Another factor that can be helpful in creating an audit trail is the practice of maintaining a log of key data processing activity. A "program run log" is a chronological record of the significant or pivotal programs that were run.

Date	Program	Type	Comments	Results

Date	- Date of activity
Program	- Program name
Type	- Program type: T = test; P = production
Comments	- e.g., error conditions encountered, problems fixed, etc.
Results	- e.g., success or failure

While it may appear that keeping a program log is a burdensome clerical task, it can be a valuable source of historical information about the sequence in which critical events occurred. It's often difficult to remember days or weeks after running a program the reasons why it was run or the context in which that activity took place.

The program run log also provides useful information for creating a project flowchart.

Managing Files

As discussed in Chapter 3, the SAS data library offers a convenient mechanism for organizing your SAS files. The library includes a directory created automatically by SAS that is used to store information about the members of the library. Individual SAS files are associated with a SAS data library through the use of a libref (short for "SAS data library reference"). All SAS files with the same libref are members of one SAS data library. SAS does not limit the number of members in a SAS data library. When multiple SAS files are members of one library, processing those files with the CONTENTS, COPY, and DATSETS procedures is simple and convenient. That is, simply referencing the library can process more than one member at a time. Another advantage is that SAS files stored in a SAS data library are logically connected through the use of a shared common name (their libref), which makes for easier file management and tracking. Several points regarding the use of SAS data libraries are listed below:

- When a SAS file is created with the same name as an existing file of the same type in the SAS data library, the original file is deleted by default after the new file is written successfully. (The NOREPLACE system option counters the default.)
- The COPY procedure can be used to copy SAS data libraries from disk to disk, disk to tape, tape to disk, and tape to tape. It is especially designed for backups.
- The DATASETS procedure can be used to rename disk-format SAS files, delete SAS files, and systematically rename a group of functionally related files in the same SAS data library. It can also be used to rename and re-label variables in SAS data sets.

Decisions regarding the storage of SAS data libraries on disk or tape should be made after consideration of the following:

- Data stored on disk are always available, while tape processing may involve waiting for a tape to mount.
- Tape storage is less flexible than disk storage since the desired file cannot be accessed directly.
- Tapes are able to accommodate very large amounts of data, which may not be practical with disk storage.

In addition, when storing SAS data set on tape, the following should be considered:

- At any point in a SAS job, only one SAS file can be accessed on a tape. For example, two SAS data sets cannot be read from the same tape in a single DATA step. However, two or more SAS files on different tapes can be accessed at the same time.
- If an existing SAS file is written over or replaced on tape, other files after it on the tape are inaccessible.

Managing Change

I have a sign in my office that reads "Adapt or Die". Even though the slogan sounds a bit Darwinian, I think it accurately reflects the nature of data analysis and today's workplace. Analysts exist because decision makers have unanswered questions. The questions will constantly change because the environment in which the decision makers operate is dynamic. This means that change should be expected. Therefore, have a process for managing change, not just reacting to it. Being able to respond and adapt to change is essential if you are going to thrive, or even survive, as an analyst.

As an analyst it's difficult to get out ahead of change, i.e., it's very hard to anticipate what the change will be, even though we know it's quite likely to occur. Much of the change we encounter results from other people changing their minds about what they need, which impacts what we do in support of those needs. I believe that the best mindset for dealing with change is to use a change management process based on one of the "Guiding Principles" from chapter 1.

If It Isn't Written Down It Was Never Said Or Done.

Follow this principle by developing the consistent practice of documenting your initial requirements (as discussed in chapter 2) and every change in scope thereafter. When change requests surface, it's your responsibility to educate the requester about the impact of those changes on delivery time and cost of the project. Use the "Change History" section of the Output Requirements document to capture that information.

In this book we've also discussed the iterative nature of analysis. As we "peel the onion" and reveal more information, the original questions may change or new questions arise. This type of change is a natural part of the analytic process. Expect it and embrace it because it means you are doing a good job. The speed and accuracy with which you are able to respond to these situations is a function of how you have structured your programs and data as well as the level of documentation you have created and maintained along the way.

Summary - Key Points

- Staying organized is not an event. It is a way of working.
- Audit trails are made not born.
- "Manage" is a verb. It implies action.
- Anticipate Change.
- "Adapt or Die".

9. Documenting the Results

Introduction

When you have successfully navigated your way through a sea of data and answered the questions posed by your client, it's time to package up the final project documentation. If you've been equally successful in staying organized throughout the course of the project, then the level of effort necessary to prepare final project documentation will be lessened. It can also be said that strong project management skills applied during the project will also contribute positively to the quality and completeness of project documentation. In the end, project documentation is the key to your ability to replicate your results.

Getting Started

In general, comprehensive project documentation should include a record of the research questions, data analysis specifications, code, test output, production output (e.g., reports), conclusions, input files, and output files. A good starting point is to recall the "5 W's" and the "1 H" from chapter 2.

- **Who** needs this information?
- **What** questions do they want answered?
- **When** do they need the answers?
- **Where** are the data?
- **Why** is this important?

Make sure that the answer to each of the "W's" is well documented. This should be as easy as including a copy of the "Output Requirements" document that was created early in the project.

Answering the "1 H" question, i.e., "**How** did I get the answer?", requires more effort. It is also the key for ensuring that your results can be replicated.

Making Results Replicable

Why is it important to be able to replicate your results? You may never be asked to do so. But what would you do if you have to prove how you came up with a certain finding? Or how would you deal with a situation where you were asked to extend a previous study by adding additional data? Your ability to respond efficiently and accurately to these scenarios depends on the quality and completeness of your project documentation. I think it's also true that knowing your results can be replicated adds a level of confidence that is evident when presenting or defending your work. It's the kind of confidence that comes from knowing you can answer any question that might be posed.

You can replicate your results if you can answer the "How" question, which can be broken down into several components:

- What you did, e.g., flowcharts, SAS logs
- When you did it, e.g., program run log
- What result you obtained, e.g., output files and listings

Expanding this further, it is my practice to save the final version of each program in the \code\prod folder for the project. In the project's \doc folder I save the SAS log and any listings that were generated for each program.

The CONTENTS and DATASETS Procedures

Several other tools are available to assist you in documenting your work. A simple and efficient technique is to take advantage of the CONTENTS and DATASETS procedures provided by the SAS System.

The CONTENTS procedure provides information about SAS data libraries and about individual SAS files in a SAS library. It is very useful for documenting permanent SAS files. Specific information pertaining to the physical characteristics of a member includes:

- The date and time of creation.
- The name of the SAS library containing the file.
- The SAS statements that generated the data set (unless NOSOURCE is specified).
- The physical data set or file where the SAS library is stored by the operating system.
- An alphabetical list of all variables and their attributes in the data set.

The output from the CONTENTS procedure can be saved as a SAS data set by using the OUT= option in the PROC CONTENTS statement.

The DATASETS procedure is a utility with multiple capabilities for managing your SAS files. One that is particularly relevant to this discussion is through use of the AUDIT statement. The AUDIT statement initiates and controls event logging to an audit file. The audit file contains a record of modifications to a SAS data set. Each time an observation is added, deleted, or updated, information is written to the audit file about who made the modification, what was modified, and when. Audit files have the same name as the original data set but a data set type of "AUDIT". The audit trail that is formed maintains a historical record of the data that enables you to trace a piece of data from the moment it enters into the data file to the time it leaves.

The following simple example demonstrates the type of information provided by the CONTENTS and DATASETS procedures.

SAS Program

```
* create a simple SAS data set ;
data one;
   x1 = .111;
   x2 = .222;
run;

proc print data=one;
run;

* invoke the datasets procedure and initiate the audit file;
proc datasets;
   audit one;
   initiate;
run;

* view the audit file ;
proc contents data=one (type=audit);
run;

* update the SAS data set ;
proc sql;
   insert into one
   set x1=.333, x2=.444;
quit;

* print a portion of the audit file ;
proc sql;
   select x1, x2, _atopcode_, _atuserid_, _atdatetime_
   from one(type=audit);
quit;
```

The SAS program shown above creates and prints a data set with one observation and two variables.

```
          The SAS System

      Obs      x1       x2

       1     0.111    0.222
```

The audit trail is initiated in PROC DATASETS with the AUDIT statement. The audit file is read by the CONTENTS procedure using the TYPE=AUDIT option and displays the output shown below.

```
                             The SAS System

                          The CONTENTS Procedure

Data Set Name:     WORK.ONE.AUDIT              Observations:            0
Member Type:       AUDIT                       Variables:               8
Engine:            V8                          Indexes:                 0
Created:           14:55 Monday, August 4, 2003   Observation Length:   82
Last Modified:     14:55 Monday, August 4, 2003   Deleted Observations:  0
Protection:                                    Compressed:              NO
Data Set Type:     AUDIT                       Sorted:                  NO
Label:


                  -----Engine/Host Dependent Information-----

Data Set Page Size:          4096
Number of Data Set Pages:    1
First Data Page:             1
Max Obs per Page:            43
Obs in First Data Page:      0
Number of Data Set Repairs:  0
File Name:                   C:\DOCUME~1\DRBret\LOCALS~1\Temp\SAS
                             Temporary Files\_TD1640\one.sas7baud
Release Created:             8.0202M0
Host Created:                WIN_PRO


                  -----Alphabetic List of Variables and Attributes-----

   #    Variable            Type      Len      Pos      Format

   3    _ATDATETIME_        Num        8        16      DATETIME19.
   8    _ATMESSAGE_         Char       8        74
   4    _ATOBSNO_           Num        8        24
   7    _ATOPCODE_          Char       2        72
   5    _ATRETURNCODE_      Num        8        32
   6    _ATUSERID_          Char      32        40
   1    x1                  Num        8         0
   2    x2                  Num        8         8
```

Then the SAS data set is updated via PROC SQL. Several audit variables contained in the audit file are updated to reflect the transaction and printed.

		The SAS System		
x1	x2	_ATOPCODE_	_ATUSERID_	_ATDATETIME_
0.333	0.444	DA	DRBRET	04AUG2003:14:55:31

The three audit variables that were specified in the SQL procedure and printed above are:

Audit Variable	Description
'_ATOPCODE_'	Stores a code describing the type of operation.
'_ATUSERID_'	Stores the logon userid associated with a modification.
'_ATDATETIME_'	Stores the date and time of a modification.

Flowcharting

Flowcharting has been around since the early days of programming. It was typically used by programmers to specify the processing steps within a program and plan the data flows between programs. I use it primarily as a tool to capture a high level visual description of data flow between each processing component in a project. It this manner a flowchart becomes a project road map after-the-fact that identifies the relative sequence of programs, each input, and each output. See the Appendix for several examples of project flowcharts.

Keeping Track of Problems Solved

In the course of completing a project it's quite likely that you will discover a new or innovative approach for manipulating data, writing efficient code, presenting results, etc. Sometimes these arise because you've worked hard to solve a problem. In other cases it's simply stumbling across a better way of doing something. Consider these events as a deposit into your intellectual capital account. Not only are these learnings valuable to you for use in future projects, they are also enormously valuable to your organization. If you're coming up with solutions and great ideas, then your colleagues probably are as well. What would happen if you pooled all of this knowledge so that you could benefit from each other's insights? Perhaps there would be less "reinventing of the wheel", i.e., solving the same problem that someone else has already solved.

I strongly encourage you to develop and maintain an inventory of techniques, great ideas, better approaches, tips and tricks, or whatever you want to call them. Build a repository of these nuggets so that they can be shared and reused by others in your organization. Depending on the culture you work in, it may be necessary to internally promote and otherwise advertise this resource and it might be necessary to motivate people to share their knowledge, since it isn't always human nature to do so.

One method that I've had success with is to use an "Issue Log" template. I keep a few of these forms around and when I come across a new technique or solution to a problem, I take a few minutes to jot it down. It usually doesn't take much time. These logs can be stored and accessed in a variety of ways; you can even load this type of information into a searchable database.

Issue Log

Date:	Project:
Initiator Name	

1. Description:

2. Turnover:

Issue Passed To:	Date:

3. Resolution:

4. Return:

Issue Received From:	Date:

5. Classification - check all that apply:

Data	☐ Missing Values ☐ Wrong Dataset ☐ Incomplete Data ☐ Different Format ☐ Unknown Values ☐ Other
Coding	☐ Faulty Logic ☐ Used Wrong Data ☐ Formula Error ☐ Selection Error ☐ Other
Infrastructure	☐ Memory ☐ Space ☐ Other
Other	

Final Project Archiving

File archiving is a critical factor in your ability to replicate results. As an analyst, you should be prepared to validate and/or re-examine previous work. Practices such as strict adherence to naming conventions, the maintenance of production code libraries, and the separation of files under unique project accounts are all designed to organize the multitude of files associated with most projects.

When the project is complete, a copy of all data, code, and documentation files should be prepared and stored in a separate physical location. Ensure that you've accounted for each component of the project that would be necessary to reconstruct any of the project outputs. Your ability to quickly and efficiently gather these items will enable you to precisely replicate a study or reopen a research question for further examination. My practice is to copy an image of each project folder (code\prod; code\test; data\prod; data\test; and \doc) into a Microsoft® Word document to file with other project documentation.

Consider the jigsaw puzzle analogy. When all the pieces have been correctly assembled and the picture is complete (no missing pieces), accurate (each piece is correctly positioned), and valid (the picture you've assembled is identical to the puzzle box cover), make a copy of each piece and store it in a separate location so that at any point in the future you could rebuild the entire puzzle into a mirror image of the original.

Summary - Key Points

Guiding Principles - Revisited

If It Isn't Written Down It Was Never Said Or Done
Identify and Document Assumptions
Get Organized
Look For Problems Early
Seek Out Master Data and Use It
Stay Organized
Archive Data As If Your Career Depended On It
Documentation Should Pass the "Hit By a Bus" Test
Expect To Be Questioned -- Possibly Doubted
Analysis Is Iterative
Confidence Comes From Knowing You Can Replicate Your Results

Appendix – Case Studies

Case Study I

Introduction

This case study converts a Microsoft Excel workbook into a set of reports that compare per capita expense categories across business areas. The level of analysis is account within business area. This is a very simple case that will serve as an introduction to the SDAR process steps.

Defining the Questions

The requester of these reports has a single primary question to be answered: "How do per capita expenses for PC's, telephone, and remote access compare across business areas?" The requester wants a monthly report that compares per capita expense across business areas to identify which areas have high utilization of these variable cost services. For the first month, the business areas selected for comparison are Legal, Public Affairs, Tax, and Treasury. In subsequent months the business areas may change. Use the "Output Requirements" template (see Figure AI.1 on the following page) to capture this and other related information.

Getting Organized

Program Names

A Project Code of "FA" will be used as the first two characters in each program name. Since the programmer is D. Blakeley, the third and fourth characters in each program name will be "DB". The combination of Task Type, Sequence Number, and Version Number is specific to each program. In all cases the extension for SAS programs will be "sas". Therefore, the first program for this project will be "FADBT01A.sas", defined as:

    FADBT01A =   FA   = Financial Analysis project
                 DB   = D. Blakeley
                 T   = test read data
                 01   = first program in the sequence
                 A   = version A of the program

Output Files

Outputs will be prefaced with the characters of the program from which they originate. For example, the output data set from the second program ("FADBR01A.sas") will be called "FADBR01A" and will be written as FADBR01A.sas7bdat.

Creating Folders

The standard set of folders was created for this project (Figure AI.2).

Figure AI.2

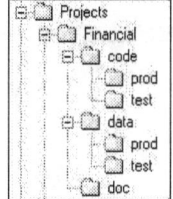

Figure AI.1

Output Requirements

Date: 8/4/2003	Requester: C. Jbara	Analyst: D. Blakeley	
Project: Financial Analysis	Report Name: Per Capita Expense Comparison		Rpt. # 001
Business Purpose	Quantify per capita expense for highly discretionary accounts to identify high utilization business areas.		
Question(s) to be Answered	"How do per capita expenses for PC's, telephone, and remote access compare across business units?"		

1. Definitions:

Inputs	1. YTD actual and budgeted expense by account within business area (Excel workbook that is extracted monthly from the G/L application). 2. Business area headcount (also contained in the same Excel workbook).
Assumptions	1. Headcount and budget data are correct.
Definitions	1. PC expense < $5,000 is account 542002. 2. Telephone expense is account 901485. 3. Remote access expense is account 901488. 4. The business areas of interest are Legal, Public Affairs, Tax, and Treasury but are subject to change in future months.
Calculations	Per Capita Expense = YTD Actual Expense / business area headcount.
Outputs	Produce a one-page hardcopy report for each account.

2. Output Format:

Title	Financial Analysis Project - Per Capita Expense Comparison - Account = xxxxxx
Header	None
Col. Headings	Y axis displays a $ scale
Row Content	X axis displays Per Capita value
Totals/Sub-Totals	n/a
Footer	Legend that identifies the business area
Graphics	Color Coded Histogram
Other	Format hours with comma and cost as whole U.S. dollar currency
Orientation: ☒ Portrait ☐ Landscape	Example Attached: ☐ Yes ☒ No

3. Output Logistics:

Medium	☒ Hardcopy	☐ Data	☐ HTML	☐ Other
Frequency	☐ Daily	☒ Monthly	☐ Quarterly	☐ Other
Type	☒ Scheduled	☐ Ad Hoc	☐ User Run	☐ Other
Delivery Location: Hand deliver to C. Jbara				
Security/Access Issues: None				

4. Agreement to Proceed:

Estimated Effort: 8 hrs.	Due Date: 8/6/2003
Requester Signature: C. Jbara	Date: Aug. 4, 2003

5. Change History:

Date	Requester	Change	Est. Effort	Due Date

Acceptance: _____   Date: _____

Creating SAS Libraries

SAS libraries for this project will follow this structure and will be assigned as follows:

Program Code	Production	'c:\Projects\Financial\code\prod'
	Test	'c:\Projects\Financial\code\test'
Data Sets	Production	'c:\Projects\Financial\data\prod'
	Test	'c:\Projects\Financial\data\test'

Gathering the Data

The data for this project are provided from the internal finance organization. The Requester, in this case the programmer, simply needs to contact another resource within the organization to obtain a copy of the Excel workbook. The request is documented by completing the "Input Requirements" template (Figure AI.3)

When the extract is received and loaded, the Requester updates the "Receipt Information" section of the "Input Requirements" document.

Creating the Analysis Dataset(s)

Reading the Data - Phase I

FADBT01A.sas

The first program tests our ability to read and convert the Excel workbook into a SAS data set.

```
*--------------------------------------------------------------------*
|Program:     FADBT01A.sas                                           |
|Source:      c:\Projects\Financial\code\test                        |
|Purpose:     Read specified sheet in an Excel workbook.             |
|Input:       2002ITSUMMARY.xls                                      |
|Output:      FADBT01A (temporary data set)                          |
|Update Log:  Notes                          Date         Programmer |
|             ------------------------------  -----------  ----------- |
|             Original version                06/09/2003   D. Blakeley |
|Usage Notes:                                                        |
|Other Notes:                                                        |
*--------------------------------------------------------------------*;

title1 'Financial Analysis Project';
title2 'Input Data Set - 2002ITSUMMARY.xls';
title3 'Data for Legal';

* read the Legal sheet in the Excel file ;
proc import
   datafile='c:\Projects\Financial\data\prod\2002ITSUMMARY.xls'
   out=FADBT01A
   dbms=excel2000
   replace;
   sheet=Legal;
run;

proc print data=FADBT01A;
run;

proc contents data=FADBT01A;
run;
```

Figure AI.3

Input Requirements

Date: 8/4/2003	Project: Financial Analysis		
Nature of Request	Please provide a copy of the Excel workbook containing the G/L extract of YTD expense and budget by account within Business Area.		
Date Required: 8/5/2003		Rush: ☒ Yes ☐ No	

1. Requester Contact:

Name	D. Blakeley	
Organization	IT Finance	
Postal Address	n/a	
E-mail Address	d.blakeley@anyorg.com	
Telephone: ext. 7826		Fax: n/a

2. Source Business Contact:

Name	Same as #3 below.	
Organization		
Postal Address		
E-mail Address		
Telephone:		Fax:

3. Source Technical Contact:

Name	R. Carr	
Organization	Corporate Finance	
Postal Address	n/a	
E-mail Address	r.carr@anyorg.com	
Telephone: ext. 2503		Fax: n/a

4. Data Content/Format:

File Type: ☐ Fixed Field ☐ Free-Format ☐ Hierarchical ☒ Other - Excel workbook		
Delimiter info.		
Time Period	Jan. 1, 2003 through Jul. 31, 2003	
Data Volume	Size: unknown	Records: unknown
Documentation: ☐ Record Layout ☐ Data Model ☐ Data Dictionary ☐ Other		
Medium: ☐ Tape ☐ CD ☐ Hardcopy ☒ Other - send via email		

5. Send To:

Name	D. Blakeley
Organization	IT Finance
Delivery Address	n/a

6. Ship Via:

U.S. Mail	☐ First Class	☐ Parcel Post	
UPS	☐ Next Day	☐ 2nd Day	☐ Ground
FedEx	☐ Next Day	☐ 2nd Day	☐ Saturday Delivery
Other	Email		

7. Receipt Information:

Date Received		Received By	
Date Loaded		Loaded By	
Filename(s)		File Location(s)	
Date Reviewed		Reviewed By	

After this program runs, we note that there are no error or warning messages in the SAS log. It appears that we are correctly converting the Excel workbook, except for a minor formatting issue. A listing of the output SAS data, a portion of which is displayed below, is compared to the Excel workbook to make sure that the conversion was accurate and complete. A review of the listing shows that the dollar values converted as dollars and cents rather than whole dollars, which indicates that the Excel file was formatted to display rounded or truncated values.

```
                            Financial Analysis Project
                         Input Data Set - 2002ITSUMMARY.xls
                                   Data for Legal

                                                                              YTD_
  Obs    Description                    Account    Headcount    YTD_Actual    Budget

   1     Total by Account - S000855         .          115          .           .
   2     Desktop SW                      542001         .        6240.15        0.0
   3     PC Expense < $5000              542002         .      104777.40    78750.0
   4     Supplies Computer               542003         .        6455.40        0.0
   5     Repairs Computer                552017         .           .           0.0
   6     Repair Parts – PCs              552018         .           .           0.0
```

Dollar fields will be rounded to the nearest integer for consistency in the next program. We've concluded that we can move out of test into production and create a permanent SAS data set.

FADBR01A.sas

This program is the production version of the code that reads the Excel workbook. It has been modified from the test version to round the dollar fields (YTD_Actual and YTD_Budget) to the nearest integer and to output a permanent SAS data set.

```
*-----------------------------------------------------------------------------*
|                                                                             |
| Program:      FADBR01A.sas                                                  |
|                                                                             |
| Source:       c:\Projects\Financial\code\prod                              |
|                                                                             |
| Purpose:      Read specified sheet in an Excel workbook.                    |
|                                                                             |
| Input:        2002ITSUMMARY.xls                                             |
|                                                                             |
| Output:       FADBR01A.sas7bdat                                             |
|                                                                             |
| Update Log:   Notes                          Date        Programmer         |
|               ----------------------------   ----------  ---------------    |
|               Original version               06/09/2003  D. Blakeley        |
|                                                                             |
| Usage Notes:                                                                |
|                                                                             |
| Other Notes:                                                                |
|                                                                             |
*-----------------------------------------------------------------------------*;
title1 'Financial Analysis Project';
title2 'Input Data Set - 2002ITSUMMARY.xls';
title3 'Data for Legal';

libname sasout 'c:\Projects\Financial\data\prod';
run;

* read the Legal sheet in the Excel file ;
proc import
   datafile='c:\Projects\Financial\data\prod\2002ITSUMMARY.xls'
   out=FADBR01A
   dbms=excel2000
   replace;
   sheet=Legal;
run;

data sasout.FADBR01A;
   set FADBR01A;
   * round off to nearest integer ;
   YTD_Actual = round(YTD_Actual);
   YTD_Budget = round(YTD_Budget);
run;

proc print data=sasout.FADBR01A;
run;

proc contents data=sasout.FADBR01A;
run;
```

We were expecting this program to run flawlessly, but a review of the SAS log reveals the following message.

```
79    data sasout.FADBR01A;
80      set FADBR01A;
81      * round off to nearest integer ;
82      YTD_Actual = round(YTD_Actual);
83      YTD_Budget = round(YTD_Budget);
84    run;

NOTE: Missing values were generated as a result of performing an operation on missing values.
      Each place is given by: (Number of times) at (Line):(Column).
      28 at 82:16    2 at 83:16
```

This is more irritating than serious. The missing values occur if the value of YTD_Actual or YTD_Budget is missing before application of the ROUND function. The value of the data does not change. However, we should have known this was going to happen when we observed missing values in the listing produced by FADBT01A.sas. Logic will be added to the next program to preclude the reoccurrence of this message in the log.

FADBR01B.sas

We've incremented this program name to the "B" version based on the addition of logic to correct the missing value issue. We've also added a macro variable that will contain the business area cost center for display on report output. Notice that the TITLE3 statement that contains the macro (&cstctr) is placed <u>after</u> the DATA step where the macro variable is created, otherwise it would not resolve.

```
*-----------------------------------------------------------------*
|                                                                 |
| Program:      FADBR01B.sas                                      |
|                                                                 |
| Source:       c:\Projects\Financial\code\prod                   |
|                                                                 |
| Purpose:      Read specified sheet in an Excel workbook.        |
|                                                                 |
| Input:        2002ITSUMMARY.xls                                 |
|                                                                 |
| Output:       FADBR01B.sas7bdat                                 |
|                                                                 |
| Update Log:   Notes                        Date         Programmer        |
|               --------------------------   -----------  ----------------- |
|               Original version             06/09/2003   D. Blakeley       |
|               Added logic for missing values  06/09/2003   D. Blakeley    |
|                                                                 |
| Usage Notes:                                                    |
|                                                                 |
| Other Notes:                                                    |
|                                                                 |
*-----------------------------------------------------------------*;

title1 'Financial Analysis Project';
title2 'Input Data Set - 2002ITSUMMARY.xls';
libname sasout 'c:\Projects\Financial\data\prod';
run;

* read the Legal sheet in the Excel file ;
proc import
   datafile='c:\Projects\Financial\data\prod\2002ITSUMMARY.xls'
   out=FADBR01B
   dbms=excel2000
   replace;
   sheet=Legal;
run;

data sasout.FADBR01B;
   set FADBR01B;
   * substring cost center from the first observation and assign to a macro variable ;
   if _n_ = 1 then do;
     call symput('cstctr',substr(description,20,7));
   end;
   * round off to nearest integer ;
```

```
   if YTD_Actual > 0 then do;
     YTD_Actual = round(YTD_Actual);
   end;
   if YTD_Budget > 0 then do;
     YTD_Budget = round(YTD_Budget);
   end;
run;

title3 "Data for Legal - Cost Center &cstctr";
proc print data=sasout.FADBR01B;
run;

proc contents data=sasout.FADBR01B;
run;
```

This program executes cleanly and produces the following listing from the CONTENTS procedure, documenting the output SAS data set.

```
                     Financial Analysis Project
                  Input Data Set - 2002ITSUMMARY.xls
                  Data for Legal - Cost Center S000855

                        The CONTENTS Procedure

Data Set Name: SASOUT.FADBR01B              Observations:         48
Member Type:   DATA                         Variables:            5
Engine:        V8                           Indexes:              0
Created:       11:10 Tuesday, August 5, 2003   Observation Length:   64
Last Modified: 11:10 Tuesday, August 5, 2003   Deleted Observations: 0
Protection:                                 Compressed:           NO
Data Set Type:                              Sorted:               NO
Label:


                -----Engine/Host Dependent Information-----

Data Set Page Size:          8192
Number of Data Set Pages:    1
First Data Page:             1
Max Obs per Page:            127
Obs in First Data Page:      48
Number of Data Set Repairs:  0
File Name:                   c:\Projects\Financial\data\prod\fadbr01b.sas7bdat
Release Created:             8.0202M0
Host Created:                WIN_PRO


                -----Alphabetic List of Variables and Attributes-----

#   Variable      Type   Len   Pos   Format   Informat   Label
----------------------------------------------------------------------------
2   Account       Num    8     0                         Account
1   Description   Char   30    32    $30.     $30.       Description
3   Headcount     Num    8     8                         Headcount
4   YTD_Actual    Num    8     16                        YTD Actual
5   YTD_Budget    Num    8     24                        YTD Budget
```

Validating the Data

The output from FADBR01B.sas should be compared in detail to the Excel workbook to ensure that the data conversion was complete and accurate. In this simple example a straightforward report-to-report comparison is sufficient.

FADBD01A.sas

At this point we're feeling comfortable with the quality of the data and are ready to calculate some per capita statistics for initial review. We are interested to see if the initial statistics are "in the ballpark" of what the requester would expect to see. Statistics for the Legal area are calculated for the three accounts specified by the Requester.

```
*--------------------------------------------------------------*
| Program:     FADBD01A.sas                                     |
|                                                               |
| Source:      c:\Projects\Financial\code\prod                  |
|                                                               |
| Purpose:     Calculate per capita statistics for specified accounts. |
|                                                               |
| Input:       FADBR01B.sas7bdat                                |
|                                                               |
| Output:      FADBD01A.sas7bdat                                |
|                                                               |
| Update Log:  Notes                            Date        Programmer        |
|              ------------------------------   ----------  ----------------- |
|              Original version                 06/09/2003  D. Blakeley       |
|                                                               |
| Usage Notes:                                                  |
|                                                               |
| Other Notes:                                                  |
|                                                               |
*--------------------------------------------------------------*;

title1 'Financial Analysis Project';
title2 'Data for Legal';

libname sasin 'c:\Projects\Financial\data\prod';
run;

data sasin.FADBD01A (drop=headcount);
  set sasin.FADBR01B;
  * create two new variables ;
  length Costcenter $ 7;
  retain Costcenter Heads;
  if _n_ = 1 then do;
    Costcenter = substr(description,20,7);
    Heads = Headcount;
  end;
  * keep accounts of interest ;
  if account in (542002,901485,901488) then do;
    * calculate per capita costs ;
    Percapita = YTD_Actual/heads;
    output;
  end;
run;

proc print data=sasin.FADBD01A;
  format  YTD_Actual
          YTD_Budget
          percapita dollar9.;
run;
```

The results are displayed below

```
                          Financial Analysis Project
                               Data for Legal

                                          YTD_       YTD_
Obs  Description              Account    Actual     Budget  Costcenter  Heads  Percapita

 1   PC Expense < $5000        542002  $104,777   $78,750   S000855      115      $911
 2   IT - Telecom              901485  $386,418  $292,950   S000855      115    $3,360
 3   IT - WW Remote Access     901488   $37,309   $31,500   S000855      115      $324
```

After reviewing these initial results with the Requester and passing the "reality check", we are ready to move on and process data for all four business areas.

Reading the Data - Phase II

FADBR02A.sas

We're now ready to convert the remaining Excel sheets for the other business areas. Rather than create a separate program for each business area, we take much of the code from FADBR01B.sas and transform it into a macro that can be called four times to separately process each business area.

The "readexcel" macro shown below appears very similar to FADBR01B.sas but contains several macro variables which enables the program to be more "data driven". Note that this code will process data from <u>any</u> business area, not just Legal.

```
%macro readexcel(dsn, businessarea);

%*-------------------------------------------------------------------*
|
| Program:      FADBR02A.mac
|
| Source:       c:\Projects\Financial\code\prod
|
| Purpose:      Read specified sheets in an Excel workbook.
|
| Input:        Specified in FADBR02A.sas
|
| Output:       FADBR02A_&businessarea
|
| Update Log:   Notes                          Date         Programmer
|               ----------------------------   -----------  ----------------
|               Original version               06/09/2003   D. Blakeley
|
| Usage Notes:  This macro can be used to process multiple business areas.
|               It is called by FADBR02A.sas.
|
| Other Notes:
|
*-------------------------------------------------------------------*;

title1 'Financial Analysis Project';
title2 "Input Data Set - &dsn";
title3 "Data for &businessarea";

%* read the business area sheet in the Excel file ;
proc import
   datafile="c:\Projects\Financial\data\prod\&dsn"
   out=FADBR02A
   dbms=excel2000
   replace;
   sheet=&businessarea;
run;

data sasout.FADBR02A_&businessarea;
   set FADBR02A;
   %* round off to nearest integer ;
   if YTD_Actual > 0 then do;
     YTD_Actual = round(YTD_Actual);
   end;
   if YTD_Budget > 0 then do;
     YTD_Budget = round(YTD_Budget);
   end;
run;

proc print data=sasout.FADBR02A_&businessarea;
run;

proc contents data=sasout.FADBR02A_&businessarea;
run;

%mend readexcel;
```

This macro is called by FADBR02A.sas; once per business area.

```
*-------------------------------------------------------------------*
|
| Program:      FADBR02A.sas
|
| Source:       c:\Projects\Financial\code\prod
|
| Purpose:      Read specified sheets in an Excel workbook.
|
| Input:        FADBR02A.mac
|               Excel workbook is specified as a parameter in macro readexcel.
|
| Output:       FADBR02A_&businessarea
|
| Update Log:   Notes                          Date         Programmer
```

```
                    ------------------------------  -----------  ------------------
                    Original version                06/09/2003   D. Blakeley
  Usage Notes: Macro readexcel can be used to process multiple business areas.
               Specify the name of the Excel workbook as the first parameter
               in macro readexcel.
               Specify the business area sheet as the second parameter.

  Other Notes:

*-----------------------------------------------------------------------*;

options mprint;

libname sasout 'c:\Projects\Financial\data\prod';
run;

%include 'c:\Projects\Financial\code\prod\FADBR02A.mac';

%readexcel(2002ITSUMMARY.xls,Legal)
%readexcel(2002ITSUMMARY.xls,PA)
%readexcel(2002ITSUMMARY.xls,Tax)
%readexcel(2002ITSUMMARY.xls,Treasury)

quit;
```

The program outputs a permanent SAS data set for each business area.

Describing the Data

FADBD02A.sas

We'll use a similar macro based approach for calculating per capita statistics for all business areas.
This program is based on FADBD01A.sas which calculated statistics for the Legal area. It is
transformed into the "calculate" macro shown below.

```
%macro calculate(businessarea);

%*-----------------------------------------------------------------------*
  Program:      FADBD02A.mac

  Source:       c:\Projects\Financial\code\prod

  Purpose:      Calculate per capita statistics for specified accounts.

  Input:        Specified in FADBD02A.sas

  Output:       FADBD02A_&businessareaa

  Update Log:   Notes                             Date         Programmer
                ------------------------------    -----------  ------------------
                Original version                  06/09/2003   D. Blakeley

  Usage Notes: This macro can be used to process multiple business areas.
               It is called by FADBD02A.sas.

  Other Notes:

*-----------------------------------------------------------------------*;

title1 'Financial Analysis Project';
title2 "Data for &businessarea";

data sasin.FADBD02A_&businessarea (drop=headcount);
  set sasin.FADBR02A_&businessarea;
  * create two new variables ;
  length Costcenter $ 7;
  retain Costcenter Heads;
  if _n_ = 1 then do;
    Costcenter = substr(description,20,7);
    Heads = Headcount;
  end;
```

```
    * keep accounts of interest ;
    if account in (542002,901485,901488) then do;
      * calculate per capita costs ;
      Percapita = YTD_Actual/heads;
      output;
    end;
  run;

  proc print data=sasin.FADBD02A_&businessarea;
    format YTD_Actual
           YTD_Budget
           percapita dollar9.;
  run;

  %mend calculate;
```

This macro is called by FADBD02A.sas; once per business area.

```
*-------------------------------------------------------------------*
|                                                                   |
|Program:      FADBD02A.sas                                         |
|                                                                   |
|Source:       c:\Projects\Financial\code\prod                     |
|                                                                   |
|Purpose:      Calculates per capita statistics for specified accounts. |
|                                                                   |
|Input:        FADBD02A.mac                                         |
|              The business area is specified as a parameter in macro |
|              calculate.                                            |
|                                                                   |
|Output:       FADBD02A_&businessarea                               |
|                                                                   |
|Update Log:   Notes                          Date        Programmer |
|              ------------------------------ ----------- --------------- |
|              Original version               06/09/2003  D. Blakeley |
|                                                                   |
|Usage Notes: Macro calculate can be used to process multiple business areas. |
|             Specify the name of the input SAS data set as the parameter in |
|             macro calculate.                                      |
|                                                                   |
|Other Notes:                                                       |
|                                                                   |
*-------------------------------------------------------------------*;

options mprint;

libname sasout 'c:\Projects\Financial\data\prod';
run;

%include 'c:\Projects\Financial\code\prod\FADBD02A.mac';

%calculate(Legal)
%calculate(PA)
%calculate(Treasury)
%calculate(Tax)

quit;
```

The program outputs a permanent SAS data set for each business area containing the information displayed below.

```
                          Financial Analysis Project
                              Data for Legal

                                    YTD_       YTD_
Obs  Description          Account  Actual     Budget  Costcenter  Heads  Percapita

1    PC Expense < $5000   542002   $104,777   $78,750   S000855     115      $911
2    IT - Telecom         901485   $386,418  $292,950   S000855     115    $3,360
3    IT - WW Remote Access 901488   $37,309   $31,500   S000855     115      $324
```

```
                          Financial Analysis Project
                                 Data for PA

                                    YTD_         YTD_
Obs  Description              Account  Actual      Budget  Costcenter  Heads  Percapita

1    PC Expense < $5000       542002   $40,520     $78,750  S100858     30     $1,351
2    IT - Telecom             901485   $160,485    $238,350 S100858     30     $5,350
3    IT - WW Remote Access    901488   $26,979     $315,000 S100858     30       $899


                          Financial Analysis Project
                              Data for Treasury

                                    YTD_         YTD_
Obs  Description              Account  Actual      Budget  Costcenter  Heads  Percapita

1    PC Expense < $5000       542002   $11,433     $26,250  S200878     17       $673
2    IT - Telecom             901485   $56,975     $70,350  S200878     17     $3,351
3    IT - WW Remote Access    901488   $3,303      $10,500  S200878     17       $194


                          Financial Analysis Project
                                 Data for Tax

                                    YTD_         YTD_
Obs  Description              Account  Actual      Budget  Costcenter  Heads  Percapita

1    PC Expense < $5000       542002   $31,950     $52,500  S200877     36       $888
2    IT - Telecom             901485   $115,563    $136,500 S200877     36     $3,210
3    IT - WW Remote Access    901488   $8,976      $10,500  S200877     36       $249
```

Answering the Questions

Manipulating the Data

<u>FADBC01A.sas</u>

Since the final report must compare per capita expenses across business areas, it is necessary to combine the statistics computed in the previous program. This is a simple step to concatenate the four statistical data sets prior to reporting.

```
*--------------------------------------------------------------------------*
| Program:      FADBC01A.sas                                               |
|                                                                          |
| Source:       c:\Projects\Financial\code\prod                            |
|                                                                          |
| Purpose:      Concatenate multiple business areas.                       |
|                                                                          |
| Input:        FADBD02A_Legal.sas7bdat                                    |
|               FADBD02A_PA.sas7bdat                                       |
|               FADBD02A_Tax.sas7bdat                                      |
|               FADBD02A_Treasury.sas7bdat                                 |
|                                                                          |
| Output:       FADBC01A.sas7bdat                                          |
|                                                                          |
| Update Log:   Notes                          Date         Programmer     |
|               ----------------------------   ----------   -----------    |
|               Original version               06/09/2003   D. Blakeley    |
|                                                                          |
| Usage Notes:                                                             |
|                                                                          |
| Other Notes:                                                             |
|                                                                          |
*--------------------------------------------------------------------------*;

title1 'Financial Analysis Project';
title2 'Combined Business Areas';

libname sasin 'c:\Projects\Financial\data\prod';
run;
```

```
* concatenate multiple business areas  ;
data sasin.FADBC01A;
   set sasin.FADBD02A_Legal
       sasin.FADBD02A_PA
       sasin.FADBD02A_Tax
       sasin.FADBD02A_Treasury
       ;
run;

proc print data=sasin.FADBC01A;
run;

proc contents data=sasin.FADBC01A;
run;
```

The output from this step is displayed below.

```
                          Financial Analysis Project
                           Combined Business Areas

                                        YTD_     YTD_
    Obs  Description            Account Actual   Budget  Costcenter  Heads Percapita

     1   PC Expense < $5000     542002  104777   78750   S000855      115    911.10
     2   IT - Telecom           901485  386418  292950   S000855      115   3360.16
     3   IT - WW Remote Access  901488   37309   31500   S000855      115    324.43
     4   PC Expense < $5000     542002   40520   78750   S100858       30   1350.67
     5   IT - Telecom           901485  160485  238350   S100858       30   5349.50
     6   IT - WW Remote Access  901488   26979  315000   S100858       30    899.30
     7   PC Expense < $5000     542002   31950   52500   S200877       36    887.50
     8   IT - Telecom           901485  115563  136500   S200877       36   3210.08
     9   IT - WW Remote Access  901488    8976   10500   S200877       36    249.33
    10   PC Expense < $5000     542002   11433   26250   S200878       17    672.53
    11   IT - Telecom           901485   56975   70350   S200878       17   3351.47
    12   IT - WW Remote Access  901488    3303   10500   S200878       17    194.29
```

In this "Analysis Dataset" we have all the elements necessary to deliver the reports as specified.

Reporting

FADBP01A.sas

The final program uses the GCHART procedure to plot the per capita expense statistics for each business area by account.

```
*-------------------------------------------------------------------------*
|                                                                         |
| Program:     FADBP01A.sas                                               |
|                                                                         |
| Source:      c:\Projects\Financial\code\prod                            |
|                                                                         |
| Purpose:     Print histogram that compares per capita expenses across   |
|              business areas.                                            |
|                                                                         |
| Input:       FADBC01A.sas7bdat                                          |
|                                                                         |
| Output:      listings to output window                                  |
|                                                                         |
| Update Log:  Notes                             Date         Programmer  |
|              -----------------------------     ----------   ----------- |
|              Original version                  06/09/2003   D. Blakeley  |
|                                                                         |
| Usage Notes:                                                            |
|                                                                         |
| Other Notes:                                                            |
|                                                                         |
*-------------------------------------------------------------------------*;

title1 'Financial Analysis Project';
title2 'Per Capita Expense Comparison';

libname sasin 'c:\Projects\Financial\data\prod';
run;

proc format;
```

```
  value acct 542002 = "     PC's     "
               901485 = '  Telephone  '
               901488 = 'Remote Access';
  value $barea 'S000855' = ' Legal  '
               'S100858' = '   PA   '
               'S200877' = '  Tax   '
               'S200878' = 'Treasury';
run;

proc sort data=sasin.FADBC01A out=sorted;
  by account;

proc gchart data=sorted;
   by account;
   vbar percapita /discrete sumvar=percapita subgroup=costcenter;
   format percapita dollar9.
          account acct.
          costcenter $barea.;
run;

quit;
```

The histogram output is displayed in Figures AI.4-6.

Figure AI.4

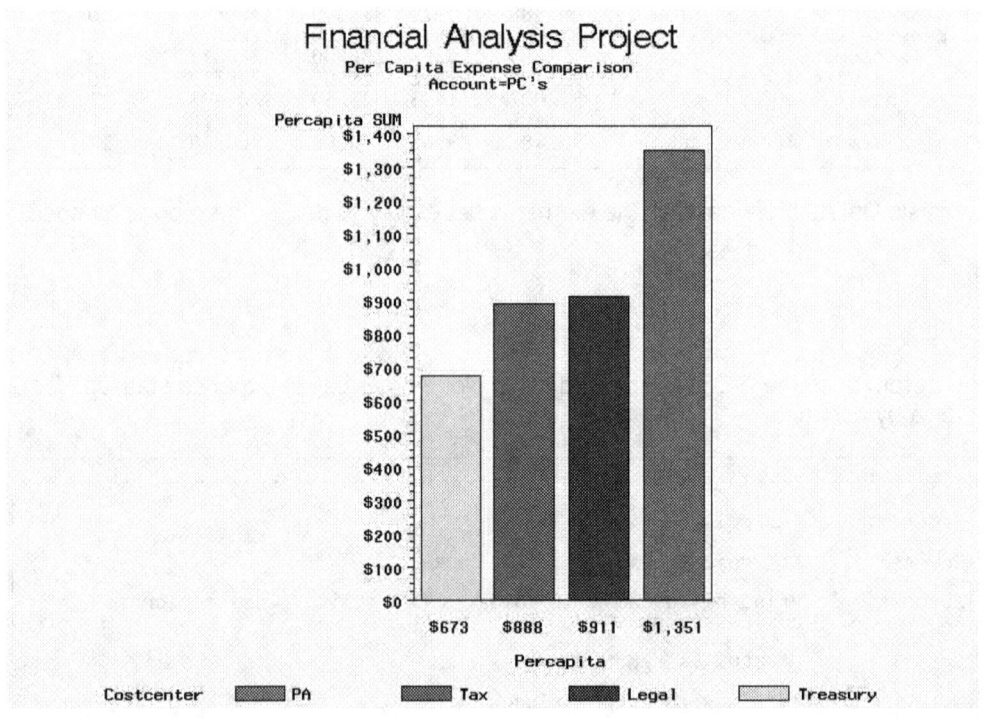

Figure AI.5

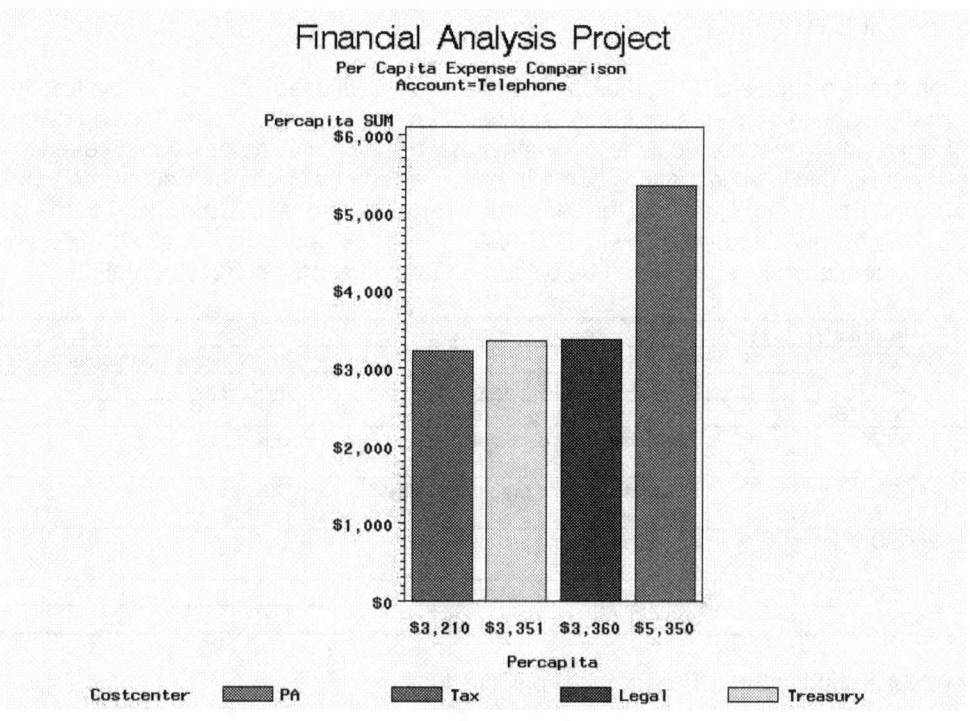

Figure AI.6

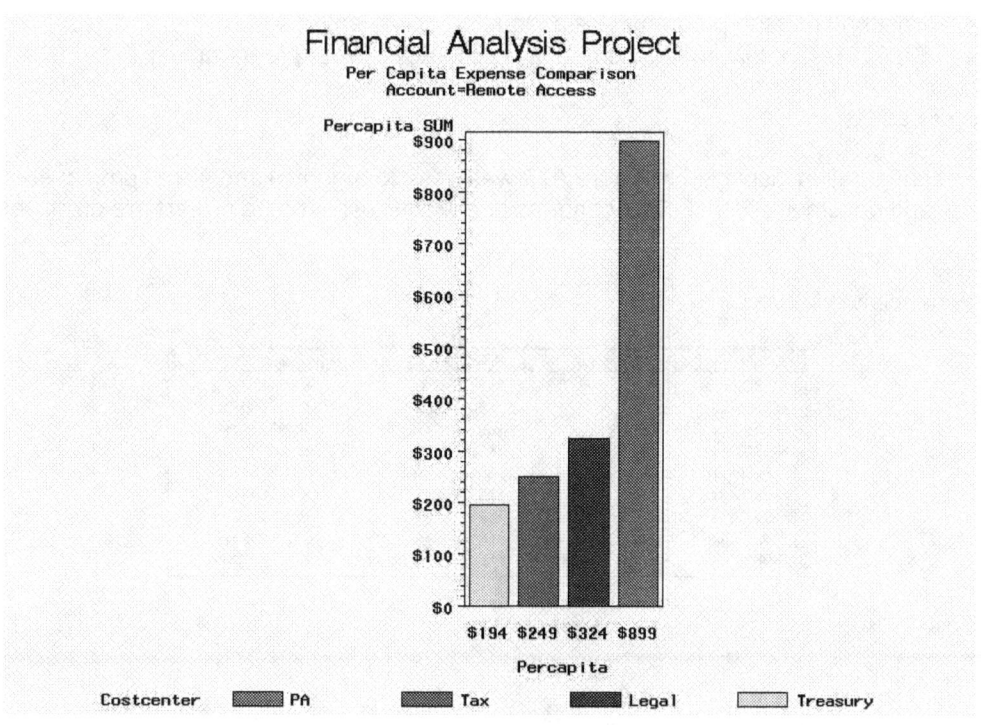

Staying Organized

Building the Audit Trail

Even though this is a simple project, a program run log was still used. The log below lists the chronological sequence of events for the most recent run of this program set. The log distinguishes test runs from production runs, includes comments that are helpful in reconstructing events, and otherwise captures information that can come in handy when sorting through output and other documentation. For example, the FADBR01B program, which processed the Excel data for the Legal area, was eventually replaced by FADBR02A. Therefore, any output that was associated with FADBR01B becomes irrelevant and can be discarded from final project documentation.

Date	Program	Type	Comments	Results
8/5/03	FADBT01A	T		Format issues with dollar fields
8/5/03	FADBR01A	P	Reads the Legal sheet	Missing values
8/5/03	FADBR01B	P	Fixes the missing values	O.K.
8/5/03	FADBD01A	P	Calculates per capita stats for review with Requester	O.K.
8/5/03	FADBR02A	P	Uses a macro to process multiple business areas	O.K.
8/5/03	FADBD02A	P	Uses a macro to process multiple business areas	O.K.
8/5/03	FADBC01A	P	Combine data for all bus. areas	O.K.
8/5/03	FADBP01A	P	plot	O.K.

Other steps taken during the course of the project include:

- Saving interim steps in the form of data sets or report listings. See the flowchart below.
- Updating the documentation block for each program to accurately reflect inputs, outputs, and changes.
- Ensuring that the programs are adequately commented.
- Saving test results that document the accuracy of your programming.

Managing Files

The folder configuration displayed in Figure AI.2 was used to organize and file all project-specific code, data, and documentation. Folder contents at the conclusion of the project are displayed below.

Projects\Financial\code\prod

Name	Size	Type	Modified
FADBC01A.sas	3 KB	SAS File	8/5/03 11:31 AM
FADBD01A.sas	3 KB	SAS File	8/5/03 2:40 PM
FADBD02A.mac	3 KB	MAC File	6/9/03 2:29 PM
FADBD02A.sas	3 KB	SAS File	8/5/03 11:26 AM
FADBP01A.sas	3 KB	SAS File	8/5/03 11:36 AM
FADBR01A.sas	3 KB	SAS File	6/9/03 10:34 AM
FADBR01B.sas	3 KB	SAS File	8/5/03 11:10 AM
FADBR02A.mac	3 KB	MAC File	6/9/03 1:46 PM
FADBR02A.sas	3 KB	SAS File	8/5/03 11:20 AM

Projects\Financial\code\test

Name △	Size	Type	Modified
FADBT01A.sas	3 KB	SAS File	8/5/03 10:52 AM
Program Doc Template.txt	2 KB	Text Document	5/13/03 2:49 PM

Projects\Financial\data\prod

Name △	Size	Type	Modified
2002ITSUMMARY.xls	302 KB	Microsoft Excel Wor...	8/5/03 10:27 AM
fadbc01a.sas7bdat	9 KB	SAS7BDAT File	8/5/03 11:33 AM
fadbd01a.sas7bdat	9 KB	SAS7BDAT File	8/5/03 11:15 AM
fadbd02a_legal.sas7bdat	9 KB	SAS7BDAT File	8/5/03 11:26 AM
fadbd02a_pa.sas7bdat	9 KB	SAS7BDAT File	8/5/03 11:26 AM
fadbd02a_tax.sas7bdat	9 KB	SAS7BDAT File	8/5/03 11:26 AM
fadbd02a_treasury.sas7bdat	9 KB	SAS7BDAT File	8/5/03 11:26 AM
fadbr01a.sas7bdat	9 KB	SAS7BDAT File	8/5/03 10:57 AM
fadbr01b.sas7bdat	9 KB	SAS7BDAT File	8/5/03 11:10 AM
fadbr02a_legal.sas7bdat	9 KB	SAS7BDAT File	8/5/03 11:20 AM
fadbr02a_pa.sas7bdat	9 KB	SAS7BDAT File	8/5/03 11:20 AM
fadbr02a_tax.sas7bdat	9 KB	SAS7BDAT File	8/5/03 11:20 AM
fadbr02a_treasury.sas7bdat	9 KB	SAS7BDAT File	8/5/03 11:20 AM

Projects\Financial\data\test -- No files.

Projects\Financial\doc

Name △	Size	Type	Modified
FADBC01A.log	4 KB	LOG File	8/5/03 11:33 AM
FADBC01A.lst	4 KB	LST File	8/5/03 11:34 AM
FADBD01A.log	4 KB	LOG File	8/5/03 11:15 AM
FADBD01A.lst	1 KB	LST File	8/5/03 11:16 AM
FADBD02A.log	9 KB	LOG File	8/5/03 11:27 AM
FADBD02A.lst	3 KB	LST File	8/5/03 11:27 AM
FADBP01A.log	4 KB	LOG File	8/5/03 11:37 AM
FADBP01A_PC.bmp	1,255 KB	Bitmap Image	8/5/03 11:40 AM
FADBP01A_REMOTE.bmp	1,255 KB	Bitmap Image	8/5/03 11:41 AM
FADBP01A_TELEPHONE.bmp	1,255 KB	Bitmap Image	8/5/03 11:40 AM
FADBR01A.log	4 KB	LOG File	8/5/03 10:58 AM
FADBR01A.lst	8 KB	LST File	8/5/03 10:58 AM
FADBR01B.log	4 KB	LOG File	8/5/03 11:10 AM
FADBR01B.lst	8 KB	LST File	8/5/03 11:10 AM
FADBR02A.log	10 KB	LOG File	8/5/03 11:21 AM
FADBR02A.lst	29 KB	LST File	8/5/03 11:22 AM
FADBT01A.log	3 KB	LOG File	8/5/03 10:54 AM
FADBT01A.lst	8 KB	LST File	8/5/03 10:54 AM
Financial Analysis Flow.vsd	59 KB	Microsoft Visio Drawi...	8/5/03 4:19 PM
PROD_DATA_CONTENTS.lst	26 KB	LST File	8/6/03 12:47 PM

Managing Change

There was no change in scope encountered in this project.

Documenting the Results

Making the Results Replicable

A review of the folder contents above shows that for each program we have saved the following information:
- Final code (in the code\prod folder)
- SAS Logs (in the doc folder)
- Sample prints (in the doc folder)

The CONTENTS and DATASETS Procedures

We have also used the CONTENTS procedure to document each production SAS data set using the code displayed below.

```
libname sasin 'c:\Projects\Financial\data\prod';
proc contents data=sasin._all_;
run;
```

Partial output from the CONTENTS procedure is displayed below.

Directory Listing of the data\prod Folder

```
                        The SAS System

                     The CONTENTS Procedure

                      -----Directory-----

            Libref:            SASIN
            Engine:            V8
            Physical Name: c:\Projects\Financial\data\prod
            File Name:         c:\Projects\Financial\data\prod


                                        File
        #  Name                Memtype  Size   Last Modified
        ─────────────────────────────────────────────────────
        1  FADBC01A            DATA     9216   05AUG2003:11:33:07
        2  FADBD01A            DATA     9216   05AUG2003:11:15:13
        3  FADBD02A_LEGAL      DATA     9216   05AUG2003:11:26:24
        4  FADBD02A_PA         DATA     9216   05AUG2003:11:26:24
        5  FADBD02A_TAX        DATA     9216   05AUG2003:11:26:25
        6  FADBD02A_TREASURY   DATA     9216   05AUG2003:11:26:25
        7  FADBR01A            DATA     9216   05AUG2003:10:57:41
        8  FADBR01B            DATA     9216   05AUG2003:11:10:02
        9  FADBR02A_LEGAL      DATA     9216   05AUG2003:11:20:14
       10  FADBR02A_PA         DATA     9216   05AUG2003:11:20:14
       11  FADBR02A_TAX        DATA     9216   05AUG2003:11:20:15
       12  FADBR02A_TREASURY   DATA     9216   05AUG2003:11:20:15
```

Contents of the "Analysis Dataset" FADBC01A.sas7bdat

```
                      The CONTENTS Procedure

Data Set Name: SASIN.FADBC01A              Observations:           12
Member Type:   DATA                        Variables:              7
Engine:        V8                          Indexes:                0
Created:       11:33 Tuesday, August 5, 2003   Observation Length:  80
Last Modified: 11:33 Tuesday, August 5, 2003   Deleted Observations: 0
Protection:                                Compressed:             NO
Data Set Type:                             Sorted:                 NO
Label:


                  -----Engine/Host Dependent Information-----

Data Set Page Size:         8192
Number of Data Set Pages:   1
First Data Page:            1
Max Obs per Page:           101
Obs in First Data Page:     12
Number of Data Set Repairs: 0
File Name:                  c:\Projects\Financial\data\prod\fadbc01a.sas7bdat
Release Created:            8.0202M0
Host Created:               WIN_PRO
```

```
              -----Alphabetic List of Variables and Attributes-----

 #    Variable        Type    Len    Pos    Format    Informat    Label
 2    Account         Num      8      0                           Account
 5    Costcenter      Char     7     70
 1    Description     Char    30     40    $30.      $30.          Description
 6    Heads           Num      8     24
 7    Percapita       Num      8     32
 3    YTD_Actual      Num      8      8                           YTD Actual
 4    YTD_Budget      Num      8     16                           YTD Budget
```

The CONTENTS procedure output should also be saved in the doc folder.

Flowcharting

The project flowchart was prepared to document each input, program, and output. It is displayed in Figure AI.7

Keeping Track of Problems Solved

As I've indicated several times, this project is simple and straightforward. However, several interesting things were learned in the process of working with the GCHART procedure. Those "learnings" are captured in an "Issue Log" and filed with the project documentation. Equally important is to make issue log information available to other analysts who may benefit from what was learned in the course of this project.

Final Project Archiving

All of the files contained in the project folders should be copied for storage on a separate device and/or physical location.

Figure AI.7

Financial Analysis

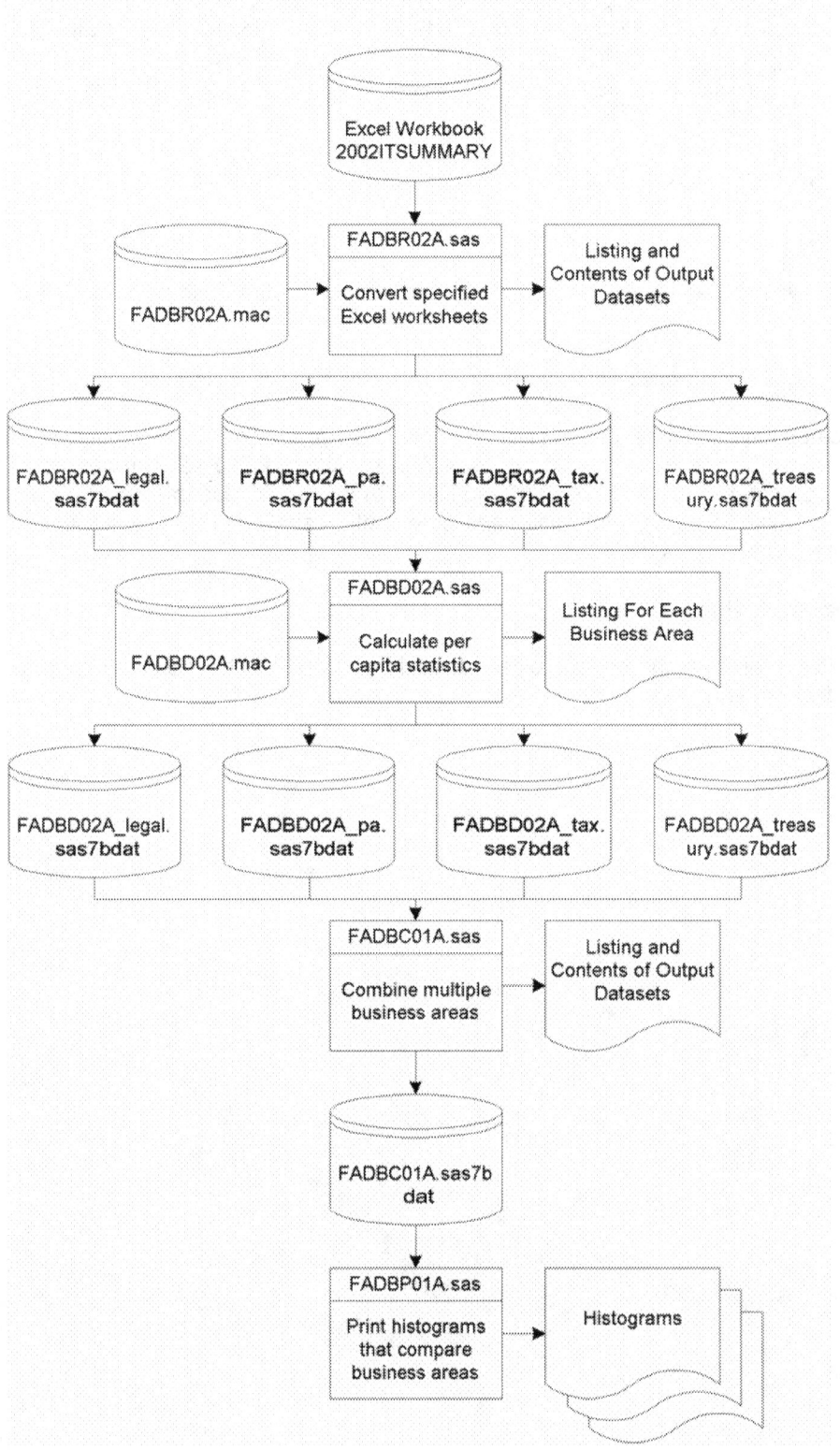

Issue Log - Working with PROC GCHART

Date: 8/6/03	Project: Financial Analysis
Initiator Name	D. Blakeley

1. Description:

I had some initial difficulty getting PROC GCHART to display data in histogram form the way I wanted. I had summary statistics (PERCAPITA) in a SAS data set that I wanted to display on the x-axis of a histogram.

The initial charts displayed frequency on the y-axis rather than dollar value; midpoint data on the x-axis rather than actual value(PERCAPITA); and only three business areas (COSTCENTER).

2. Turnover:

Issue Passed To: File	Date: 8/6/03

3. Resolution:

I found that using the following options in the VBAR statement solved the problem.
- discrete   Treats a numeric variable as a discrete variable, rather than a continuous variable. A separate midpoint is created for each unique formatted value of the chart variable.
- sumvar=   Calculates the sum of the summary-variable for each midpoint.
- subgroup=   Creates a separate segment within each bar for every unique value of the subgroup variable for that midpoint.

The VBAR statement ended up as follows:
VBAR PERCAPITA / DISCRETE SUMVAR=PERCAPITA SUBGROUP=COSTCENTER;

This issue is resolved.

4. Return:

Issue Received From: n/a	Date: n/a

5. Classification - check all that apply:

Data	☐ Missing Values	☐ Wrong Dataset	☐ Incomplete Data
	☐ Different Format	☐ Unknown Values	☐ Other
Coding	☐ Faulty Logic	☐ Used Wrong Data	☐ Formula Error
	☐ Selection Error	☒ Other - PROC GCHART VBAR statement	
Infrastructure	☐ Memory	☐ Space	☐ Other
Other			

Case Study II

Introduction

This case study draws a sample of inpatient hospital discharges from a national hospital claims source. The sample is used to calculate the change in cost-to-charge ratios over time. In the first set of reports the level of analysis is the census region. In the second set of reports the level of analysis is the Major Diagnostic Category (MDC).

Defining the Questions

The requester of these reports has a multifaceted question to be answered: "How are cost-to-charge ratios changing over time (year-to-year comparison) and how do these changes, if any, vary by geographic region and medical condition?" Use the "Output Requirements" template (see Figure AII.1 on the following page) to capture this and other related information.

Getting Organized

Program Names

A Project Code of "HC" will be used as the first two characters in each program name. Since the programmer is D. Blakeley, the third and fourth characters in each program name will be "DB". The combination of Task Type, Sequence Number, and Version Number is specific to each program. In all cases the extension for SAS programs will be "sas". Therefore, the first program for this project will be "HCDBT01A.sas", defined as:

HCDBR01A =	HC	= Health Care Cost project
	DB	= D. Blakeley
	T	= test read data
	01	= first program in the sequence
	A	= version A of the program

Output Files

Outputs will be prefaced with the characters of the program from which they originate. For example, the output data set from the second program ("HCDBR01A.sas") will be called "HCDBR01A" and will be written as HCDBR01A.sas7bdat.

Creating Folders

The standard set of folders, plus the addition of a folder for user-defined formats was created for this project (Figure AII.2).

Figure AII.2

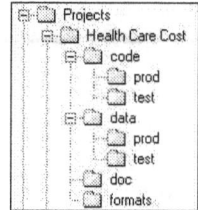

Figure AII.1

Output Requirements

Date: 5/28/2003	Requester: L. Heineccius	Analyst: D. Blakeley	
Project: Health Care Cost	Report Name: Percentage Changes		Rpt. # 001
Business Purpose	Understand the trends in inpatient costs and charges per discharge.		
Question(s) to be Answered	How are cost-to-charge ratios changing over time (year-to year comparison) and how do these changes, if any, vary by geographic region and medical condition?		

1. Definitions:

Inputs	1. 1999-2001 National Hospital Claims.
Assumptions	1. The data source contains inpatient hospital discharge data representative of calendar years 1999 through 2001.
Definitions	1. MDC = "Major Diagnostic Category". 2. Census Region = the four census regions within the United States. 3. Period 1 = Jan 00 4. Period 2 = Jan 01
Calculations	1. Calculate the average charge and cost by census region and year. 2. Calculate the average charge and cost by MDC and year. 3. Calculate the cost-to-charge ratio by census region and year, where ccratio=avg cost/avg. chg. 4. Calculate the cost-to-charge ratio by MDC and year, where ccratio=avg. cost/avg. chg. 5. Calculate the percentage change in mean charge, cost, and cost-to-charge ratio between years by census region, where change=(year2-year1)/year1. 6. Calculate the percentage change in mean charge, cost, and cost-to-charge ratio between years by MDC, where change=(year2-year1)/year1.
Outputs	Produce two hardcopy reports for census regions and MDCs: 1. Listing of percentage change in charges, costs, and cost-to-charge ratios. 2. Plot of percentage change in charges, costs, and cost-to-charge ratios.

2. Output Format:

Title	Percentage Changes by Census Region and Year; by MDC and Year		
Header	None		
Col. Headings	Census Region (or MDC), % Change Charges, % Change Costs, % Change C-C Ratio		
Row Content	1 row per census region/year; MDC/year		
Totals/Sub-Totals	None		
Footer	None		
Graphics	Plots for all the above.		
Other	None		
Orientation:	☒ Portrait ☐ Landscape	Example Attached: ☐ Yes	☒ No

3. Output Logistics:

Medium	☒ Hardcopy	☐ Data	☐ HTML	☐ Other
Frequency	☐ Daily	☐ Monthly	☐ Quarterly	☐ Other
Type	☐ Scheduled	☒ Ad Hoc	☐ User Run	☐ Other
Delivery Location: Hand deliver to L. Heineccius				
Security/Access Issues: None				

4. Agreement to Proceed:

Estimated Effort: 30 hrs.	Due Date: 6/9/2003
Requester Signature: L. Heineccius	Date: May 2, 2003

5. Change History:

Date	Requester	Change	Est. Effort	Due Date
6/4/2003	D. Blakeley	Need additional data	10 hrs.	8/7/2003

Acceptance: _____   Date: _____

Creating SAS Libraries

SAS libraries for this project will follow this structure and will be assigned as follows:

Program Code	Production	'c:\Projects\Health Care Cost\code\prod'
	Test	'c:\Projects\Health Care Cost\code\test'
Data Sets	Production	'c:\Projects\Health Care Cost\data\prod'
	Test	'c:\Projects\Health Care Cost\data\test'
User Defined Formats		'c:\Projects\Health Care Cost\formats'

Gathering the Data

The data for this project is provided by a vendor but housed internally. The requester, in this case the programmer, simply needs to contact another technical resource within the organization to request access to the Oracle table that contains patient discharge data. The request is documented by completing the "Input Requirements" template (Figure AII.3)

Since data are not being physically transferred, the "Receipt Information" section of the "Input Requirements" document is not relevant.

Creating the Analysis Dataset(s)

Reading the Data

RUDBT01A.sas
The first program takes an initial look at the patient_discharge table. The library in which the table resides has been defined using the SAS/ACCESS® interface for Oracle. This allows us to read the table as if it were a SAS data set. The program produces a PROC CONTENTS listing, prints the first five observations, and displays a frequency of discharges by month.

```
*-----------------------------------------------------------------------*
|Program:      HCDBT01A.sas
|
|Source:       c:\Projects\Health Care Cost\code\prod
|
|Purpose:      Produces a frequency of patient discharges by month from the
|              entire 1999-2001 National Hospital Claims sample.
|
|Input:        AppIII\patient_discharge (Oracle table)
|
|Output:       Reports to the output window
|
|Update Log:   Notes                           Date          Programmer
|              ------------------------------   -----------   ----------------
|              Original version                 05/29/2003    D. Blakeley
|
|Usage Notes:
|
|Other Notes:
|
*-----------------------------------------------------------------------*;
title 'Contents of Patient_Discharge Table';
proc contents data=appiii.patient_discharge;
run;

title 'Sample Obs. from Patient_Discharge Table';
proc print data=appiii.patient_discharge (obs=5);
run;

title 'Frequency of Patient Discharges by Month';
proc freq data=appiii.patient_discharge;
   tables file_year*discharge_month;
run;
```

Figure AII.3

Input Requirements

Date: 6/2/2003	Project: Health Care Cost			
Nature of Request	Access to the Oracle patient_discharge table in the National Hospital Claims database.			
Date Required: 6/3/2003		Rush: ☐ Yes	☒ No	

1. Requester Contact:

Name	D. Blakeley	
Organization	IT Finance	
Postal Address	n/a	
E-mail Address	d.blakeley@anyorg.com	
Telephone: ext. 7826		Fax: n/a

2. Source Business Contact:

Name	Same as #3 below.	
Organization		
Postal Address		
E-mail Address		
Telephone:		Fax:

3. Source Technical Contact:

Name	S. Edison	
Organization	Corporate IT	
Postal Address	n/a	
E-mail Address	s.edison@anyorg.com	
Telephone: ext. 2571		Fax: n/a

4. Data Content/Format:

File Type: ☒ Fixed Field	☐ Free-Format	☐ Hierarchical	☐ Other - variable length records
Delimiter info.			
Time Period	Calendar year 1999 through 2001		
Data Volume	Size: unknown	Records: unknown	
Documentation: ☐ Record Layout	☐ Data Model	☐ Data Dictionary ☐ Other	
Medium: ☐ Tape ☐ CD ☐ Hardcopy ☐ Other			

5. Send To:

Name	n/a
Organization	
Delivery Address	

6. Ship Via:

U.S. Mail	☐ First Class	☐ Parcel Post	
UPS	☐ Next Day	☐ 2nd Day	☐ Ground
FedEx	☐ Next Day	☐ 2nd Day	☐ Saturday Delivery
Other	Email me when access has been granted.		

7. Receipt Information:

Date Received		Received By	
Date Loaded		Loaded By	
Filename(s)		File Location(s)	
Date Reviewed		Reviewed By	

The PROC CONTENTS output provides an alphabetic list of all variables and their attributes in the table. The frequency table of discharges by month is displayed below.

```
                    Table of FILE_YEAR by DISCHARGE_MONTH

FILE_YEAR(FILE_YEAR)        DISCHARGE_MONTH(DISCHARGE_MONTH)
```

Frequency Percent Row Pct Col Pct	1	2	3	4	5	6	Total
2000	0 0.00 0.00 0.00	0 0.00 0.00 0.00	0 0.00 0.00 0.00	0 0.00 0.00 0.00	0 0.00 0.00 0.00	0 0.00 0.00 0.00	542024 15.57
2001	197776 5.68 9.08 55.88	183885 5.28 8.44 55.90	199412 5.73 9.16 55.79	184410 5.30 8.47 61.11	189525 5.44 8.70 63.64	183123 5.26 8.41 70.67	2178036 62.57
2002	156126 4.49 20.52 44.12	145083 4.17 19.07 44.10	158039 4.54 20.77 44.21	117362 3.37 15.42 38.89	108305 3.11 14.23 36.36	76018 2.18 9.99 29.33	760933 21.86
Total	353902 10.17	328968 9.45	357451 10.27	301772 8.67	297830 8.56	259141 7.44	3480993 100.00

(Continued)

Frequency Percent Row Pct Col Pct	7	8	9	10	11	12	Total
2000	0 0.00 0.00 0.00	0 0.00 0.00 0.00	0 0.00 0.00 0.00	183017 5.26 33.77 51.42	177761 5.11 32.80 51.58	181246 5.21 33.44 51.82	542024 15.57
2001	176752 5.08 8.12 100.00	184771 5.31 8.48 100.00	170104 4.89 7.81 100.00	172924 4.97 7.94 48.58	166852 4.79 7.66 48.42	168502 4.84 7.74 48.18	2178036 62.57
2002	0 0.00 0.00 0.00	0 0.00 0.00 0.00	0 0.00 0.00 0.00	0 0.00 0.00 0.00	0 0.00 0.00 0.00	0 0.00 0.00 0.00	760933 21.86
Total	176752 5.08	184771 5.31	170104 4.89	355941 10.23	344613 9.90	349748 10.05	3480993 100.00

A review of this frequency table reveals some surprising information about our data assumptions. This table contains data from October 2000 through June 2002, rather than calendar years 1999 through 2001 as we had expected. Therefore, we will need to revisit the specifications that the year-over-year comparison be based on data for January 2000 and January 2001. A review of these findings with the requester resulted in the decision to base the analysis on a comparison of data from December 2000 and December 2001.

HCDBR01A.sas

This program read the Oracle table and randomly selects discharges from December 2000 (181,246 discharges) or December 2001 (168,502 discharges). The objective is to select approximately three percent of the observations available in these time periods. This should produce a sample of approximately 10,492 observations.

```
*---------------------------------------------------------------*
|                                                               |
|Program:      HCDBR01A.sas                                     |
|                                                               |
|Source:       c:\Projects\Health Care Cost\code\prod           |
|                                                               |
|Purpose:      Reads the 1999-2001 National Hospital Claims sample
|              and randomly extracts selected variables from the |
|              patient_discharge table.                         |
|                                                               |
|Input:        AppIII\patient_discharge (Oracle table)          |
|                                                               |
|Output:       HCDBR01A.sas7bdat                                |
|              sample print to output window                    |
|                                                               |
|Update Log:   Notes                         Date        Programmer  |
|              ----------------------------  ----------  ----------- |
|              Original version              05/29/2003  D. Blakeley  |
|                                                               |
|Usage Notes:  A random number function is set to select approximately
|              3% of observations from the 3.4 million record input file. |
|                                                               |
|Other Notes:  This program reads Oracle data directly into a SAS data set |
|              using the Oracle access engine.                  |
*---------------------------------------------------------------*;
libname sasout 'c:\Projects\Health Care Cost\data\prod';
run;

data sasout.HCDBR01A;
  set appiii.patient_discharge
      (keep=file_year discharge_month mdc total_charges total_costs census_region);
    * limit extract to December discharges in 2000 or 2001 ;
    if discharge_month=12 and (file_year=2000 or file_year=2001);
    * randomly select 3 percent of observations ;
    if ranuni(0) <.03;
run;

title 'Sample Obs. from HCDBR01A';
proc print data=sasout.HCDBR01A (obs=20);
run;
```

The SAS log documents the result of the sampling. The output SAS data set containing the sample has 10,544 observations, which is very close to our estimate of 10,492.

```
NOTE: There were 3480993 observations read from the data set APPIII.PATIENT_DISCHARGE.
NOTE: The data set SASOUT.HCDBR01A has 10544 observations and 6 variables.
NOTE: DATA statement used:
      real time            11:56.57
      cpu time             41.90 seconds
```

Describing the Data

HCDBV01A.sas

This is the first validation/description program to assess the sample data set. This program produces a variety of descriptive statistics. Excerpts from the output are displayed below.

Obs	FILE_ YEAR	DISCHARGE_ MONTH	_FREQ_	Tot_Cgs	Tot_Cst	Ave_Cgs	Ave_Cst
		Summary of Total Charges and Costs by Year - HCDBR01A					
1	2000	12	5408	70722828.18	33212667.00	13077.45	6141.40
2	2001	12	5136	74672765.71	31487928.50	14539.09	6130.83

From this report we can confirm that the sample contains only discharges from December 2000 and December 2001. The total number of observations in the sample data set is distributed fairly evenly across both years. The average cost per discharge is also roughly comparable. Unfortunately, this project has no external comparison data available to otherwise validate our data set.

```
                          The UNIVARIATE Procedure
                 Variable:  TOTAL_CHARGES   (TOTAL_CHARGES)

                       Basic Statistical Measures

            Location                           Variability

      Mean      13789.42       Std Deviation            24125
      Median     7925.68       Variance             582015404
      Mode          0.00       Range                   847375
                               Interquartile Range      11146

                        Extreme Observations

         -----Lowest-----              -----Highest----

          Value       Obs              Value       Obs

        -180.00       984            416579       9103
        -132.04      4742            504236       3384
        -128.54      6311            522924       9102
           0.00      9600            624274       3867
           0.00      9595            847195       7941
```

```
                          The UNIVARIATE Procedure
                  Variable:  TOTAL_COSTS   (TOTAL_COSTS)

                       Basic Statistical Measures

            Location                           Variability

      Mean       6136.248      Std Deviation             9717
      Median     3612.140      Variance              94418971
      Mode         11.570      Range                   326010
                               Interquartile Range       4724

                        Extreme Observations

         ------Lowest-----             -----Highest----

          Value       Obs              Value       Obs

       -2213.48      8840            173624       3384
       -1490.52      8869            176081       4391
        -372.37      9707            180179       4376
         -79.75      4742            211180       7941
         -77.91      6311            323796       4387
```

These reports indicate a high degree of variability for both Charges and Costs. Unusually high and low values are observed for both variables as well. For example, the cost associated with a hospital discharge is not expected to be negative.

Other reports show the frequency of discharges by MDC and Census Region.

The FREQ Procedure				
MDC				
MDC	Frequency	Percent	Cumulative Frequency	Cumulative Percent
1	547	5.19	547	5.19
10	268	2.54	815	7.73
11	325	3.08	1140	10.81
12	69	0.65	1209	11.47
13	338	3.21	1547	14.67
14	1263	11.98	2810	26.65
15	1139	10.80	3949	37.45
16	92	0.87	4041	38.33
17	93	0.88	4134	39.21
18	197	1.87	4331	41.08
19	420	3.98	4751	45.06
2	14	0.13	4765	45.19
20	84	0.80	4849	45.99
21	134	1.27	4983	47.26
22	5	0.05	4988	47.31
23	145	1.38	5133	48.68
24	27	0.26	5160	48.94
25	28	0.27	5188	49.20
3	145	1.38	5333	50.58
4	1069	10.14	6402	60.72
5	1957	18.56	8359	79.28
6	829	7.86	9188	87.14
7	257	2.44	9445	89.58
8	886	8.40	10331	97.98
9	213	2.02	10544	100.00

CENSUS_REGION				
CENSUS_ REGION	Frequency	Percent	Cumulative Frequency	Cumulative Percent
1	255	2.42	255	2.42
2	3292	31.22	3547	33.64
3	6156	58.38	9703	92.02
4	841	7.98	10544	100.00

Another perspective on the issues of variation and extreme values is represented by the plots of Total Charges (Figure AII.4) and Total Costs (Figure AII.5).

Figure AII.4

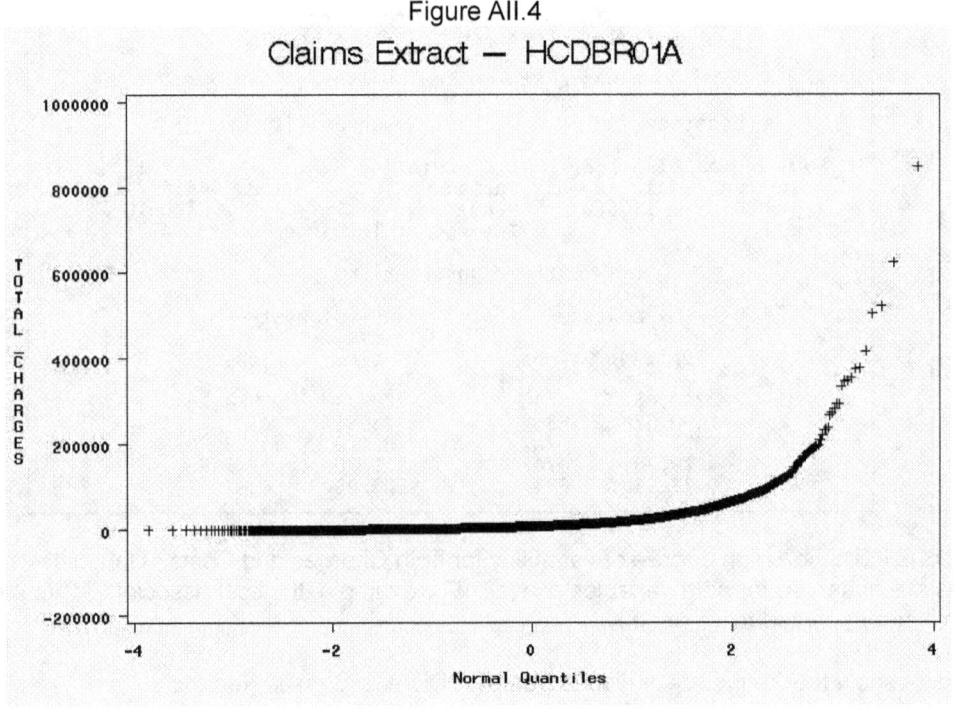

Claims Extract — HCDBR01A

Figure AII.5

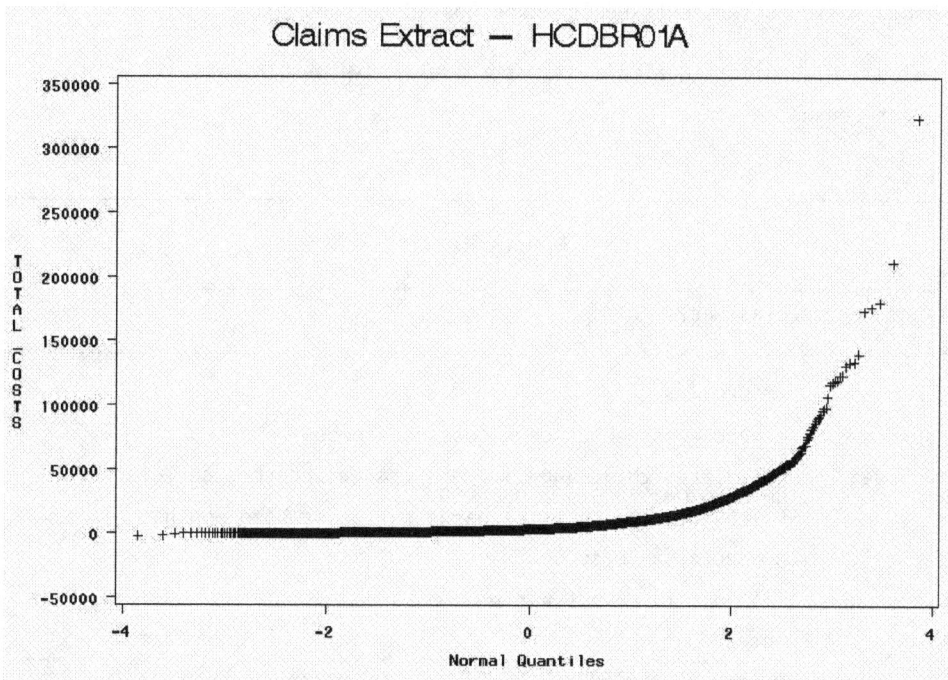

All of these reports provide evidence of high variation and the presence of both high and low outliers for Charges and Costs. In spite of these "red flags", an initial set of reports was produced for review with the requester.

Answering the Questions

HCDBP01A.sas

This program calculates and plots the percentage change in mean charge, cost, and cost-to-charge ratios between December 2000 and December 2001. A macro is used so that the same code can be run regardless of whether the BY variable is Census Region or MDC.

```
*------------------------------------------------------------------------*
| Program:      HCDBP01A.sas                                             |
|                                                                         |
| Source:       c:\Projects\Health Care Cost\code\prod                   |
|                                                                         |
| Purpose:      Plots the percentage change in mean charge, cost, and cost-
|               charge ratios between Dec 2000 and Dec 2001.             |
|               The level of analysis is the discharge.                  |
|                                                                         |
| Input:        HCDBP01A.mac                                             |
|               Data set is specified in macro report.                   |
|                                                                         |
| Output:       listings to output window                                |
|                                                                         |
| Update Log:   Notes                        Date         Programmer     |
|               ---------------------------   -----------  -------------- |
|               Original version              06/02/2003   D. Blakeley    |
|                                                                         |
| Usage Notes:  Specify the input data set name in macro report.         |
|                                                                         |
| Other Notes:                                                            |
*------------------------------------------------------------------------*;
```

```
options nodate mprint;

libname sasin 'c:\Projects\Health Care Cost\data\prod';
run;

%include 'c:\Projects\Health Care Cost\code\prod\HCDBP01A.mac';

%report(census_region)
%report(mdc)

quit;
```

Macro "Report"

```
%macro report(classvar);

%*--------------------------------------------------------------------------*

 Program:      HCDBP01A.mac

 Source:       c:\Projects\Health Care Cost\code\prod

 Purpose:      Calculates the change in mean charge and cost per discharge
               between Dec 2000 and Dec 2001.
               Calculates the cost-to-charge ratio for both months.

 Input:        HCDBR01A.sas7bdat

 Output:       listings to output window

 Update Log:   Notes                             Date          Programmer
               ------------------------------    -----------   ------------------
               Original version                  06/02/2003    D. Blakeley

 Usage Notes:

 Other Notes:

*--------------------------------------------------------------------------*;
title1 'Health Care Cost Project';
title2 'Input Data Set - HCBDR01A';
title3 'Data for Dec 2000 and Dec 2001';

%* calculate average charge and cost by (census region or mdc) and year ;
proc summary data=sasin.HCDBR01A nway;
  class &classvar file_year;
  var total_charges total_costs;
  output out=stats mean=avgchg avgcst;
run;

%* calculate the cost-to-charge ratio for each year ;
data ccratio;
  set stats;
  ccratio=avgcst/avgchg;
run;

proc sort data=ccratio out=sorted;
  by &classvar;
run;

%* calculate the percentage change in mean charge, cost, and cost-to-charge
ratio ;
data delta;
  set sorted;
  by &classvar;
  retain avgchg_d avgcst_d ccratio_d;
  if first.&classvar then do;
    avgchg_d=avgchg;
    avgcst_d=avgcst;
    ccratio_d=ccratio;
  end;
  if last.&classvar then do;
    dchg=(avgchg-avgchg_d)/avgchg_d;
    dcst=(avgcst-avgcst_d)/avgcst_d;
    dccr=(ccratio-ccratio_d)/ccratio_d;
  end;
run;


title4 "Percentage Changes by &classvar and Year";
```

118

```
proc print data=delta split='*' noobs;
  var &classvar
      file_year
      _freq_
      avgchg
      avgcst
      ccratio
      dchg
      dcst
      dccr;
  label file_year = 'Claim*Year'
        _freq_    = 'Patient*Discharges'
        avgchg    = 'Average*Charge'
        avgcst    = 'Average*Cost'
        ccratio   = 'Cost-to-Charge*Ratio'
        dchg      = '% Change*Charges'
        dcst      = '% Change*Costs'
        dccr      = '% Change*C-C Ratio';
run;

title4 'Percentage Change in Mean Charge, Cost, and Cost-to-Charge Ratios';
proc plot data=delta;
  plot dchg*&classvar;
  plot dcst*&classvar;
  plot dccr*&classvar;
run;

%mend report;
```

This program produces multiple reports that document the adverse impact of outlier observations on the calculations. Several excerpts from the MDC reports are displayed below to characterize this issue.

```
                          Health Care Cost Project
                         Input Data Set - HCBDR01A
                        Data for Dec 2000 and Dec 2001
                        Percentage Changes by mdc and Year
```

MDC	Claim Year	Patient Discharges	Average Charge	Average Cost	Cost-to-Charge Ratio	% Change Charges	% Change Costs	% Change C-C Ratio
2	2000	5	9417.03	3168.51	0.33647	.	.	.
2	2001	9	13093.72	5140.23	0.39257	0.39043	0.62229	0.16675
20	2000	50	5118.24	2681.07	0.52383	.	.	.
20	2001	34	5838.23	3332.51	0.57081	0.14067	0.24298	0.08969
21	2000	64	9699.94	5394.81	0.55617	.	.	.
21	2001	70	10622.80	4575.57	0.43073	0.09514	-0.15186	-0.22554
22	2000	4	99030.14	84708.74	0.85538	.	.	.
22	2001	1	77530.13	42146.73	0.54362	-0.21711	-0.50245	-0.36448
23	2000	69	15659.01	9470.48	0.60479	.	.	.
23	2001	76	16676.01	9288.60	0.55700	0.06495	-0.01920	-0.07902
24	2000	12	47903.32	17281.10	0.36075	.	.	.
24	2001	15	39708.04	17477.04	0.44014	-0.17108	0.01134	0.22007
25	2000	18	14481.51	6074.23	0.41945	.	.	.
25	2001	10	63181.77	21893.59	0.34652	3.36293	2.60434	-0.17387
3	2000	87	10429.95	5271.77	0.50545	.	.	.
3	2001	58	12143.03	4776.66	0.39337	0.16425	-0.09392	-0.22174

A portion of the MDC report is displayed above. It shows unusual values for "% Change Charges" for MDC 25, which has only ten discharges for December 2001. The plot of the same statistic, shown on the next page, further highlights this finding. Note that the "report" macro will need to be modified to correctly format the percentage values for both reports.

These findings send us back to the drawing board. Further analysis is required to better understand the nature of the discharge data set.

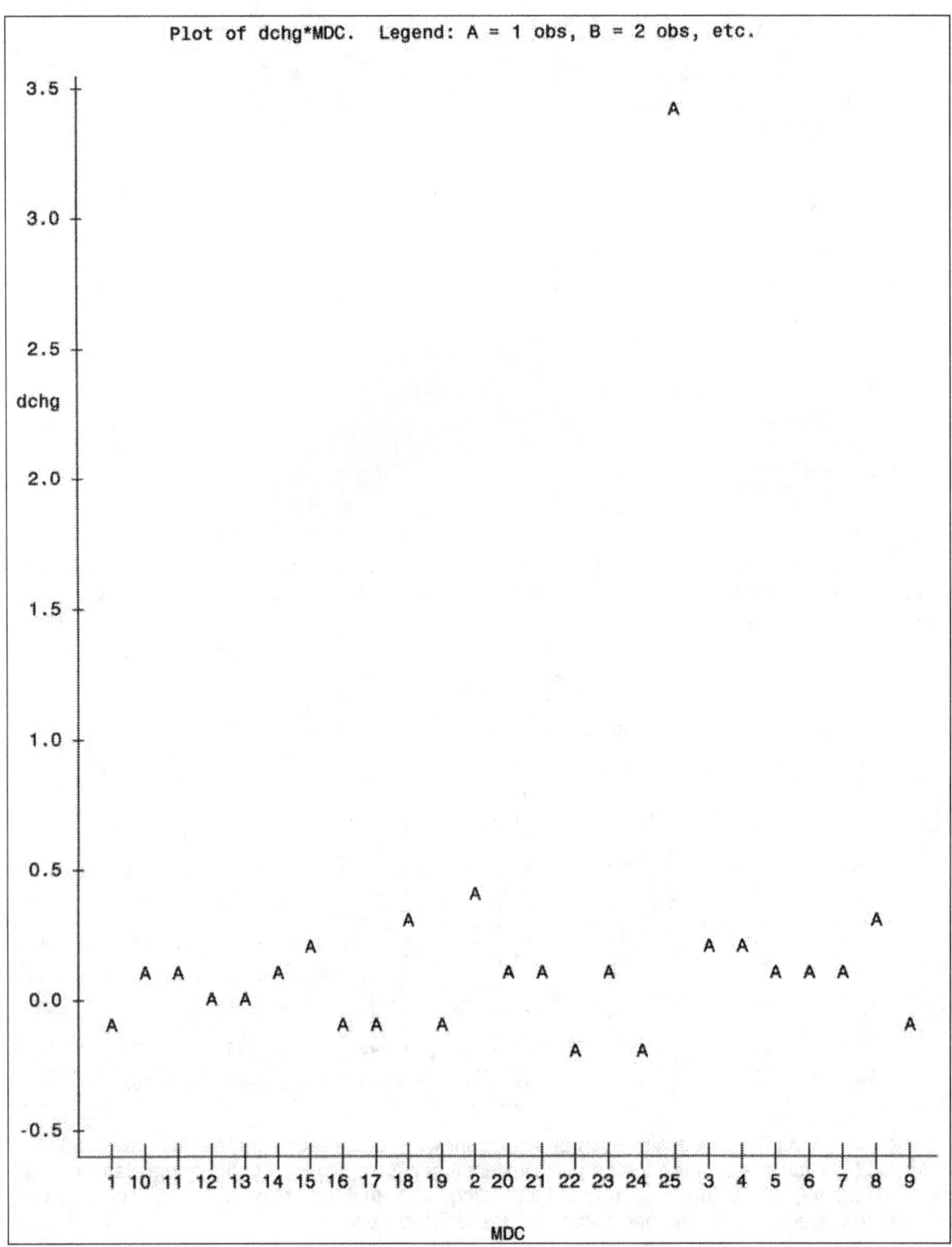

Plot of dchg*MDC.   Legend: A = 1 obs, B = 2 obs, etc.

Describing the Data (again)

The presence of outliers in the data set appears to be skewing the calculations. We need to dig deeper into this situation in order to formulate an appropriate strategy.

HCDBD01A.sas

This program classifies each discharge into one of ten Charge outlier groups and one of ten Cost outlier groups. The frequency of discharges in both sets of outlier groups is then calculated.

```
*-------------------------------------------------------------------*
|                                                                   |
| Program:     HCDBD01A.sas                                         |
|                                                                   |
| Source:      c:\Projects\Health Care Cost\code\prod               |
|                                                                   |
| Purpose:     Outlier analysis of total_charges and total_cost.    |
|                                                                   |
| Input:       HCDBR01A.sas7bdat                                    |
|                                                                   |
| Output:      Report to output window                              |
|                                                                   |
| Update Log:  Notes                       Date        Programmer   |
|              ----------------------       ----------  ----------------
|              Original version             06/04/2003  D. Blakeley |
|                                                                   |
| Usage Notes:                                                      |
|                                                                   |
| Other Notes:                                                      |
|                                                                   |
*-------------------------------------------------------------------*;

libname sasin 'c:\Projects\Health Care Cost\data\prod';
libname library 'c:\Projects\Health Care Cost\formats';
run;

data HCDBD01A;
  set sasin.HCDBR01A;
  * classify total_charge values ;
  select;
    when (              total_charges < 1     ) Chg_Level = '01';
    when (1       <= total_charges < 500   ) Chg_Level = '02';
    when (500     <= total_charges < 1000  ) Chg_Level = '03';
    when (1000    <= total_charges < 5000  ) Chg_Level = '04';
    when (5000    <= total_charges < 10000 ) Chg_Level = '05';
    when (10000   <= total_charges < 50000 ) Chg_Level = '06';
    when (50000   <= total_charges < 100000) Chg_Level = '07';
    when (100000  <= total_charges < 200000) Chg_Level = '08';
    when (200000  <= total_charges < 300000) Chg_Level = '09';
    when (300000  <= total_charges < 400000) Chg_Level = '10';
    otherwise chg_level = '99';
  end;
  * classify total_cost values ;
  select;
    when (              total_costs < 1     ) Cst_Level = '01';
    when (1       <= total_costs < 500   ) Cst_Level = '02';
    when (500     <= total_costs < 1000  ) Cst_Level = '03';
    when (1000    <= total_costs < 5000  ) Cst_Level = '04';
    when (5000    <= total_costs < 10000 ) Cst_Level = '05';
    when (10000   <= total_costs < 50000 ) Cst_Level = '06';
    when (50000   <= total_costs < 100000) Cst_Level = '07';
    when (100000  <= total_costs < 200000) Cst_Level = '08';
    when (200000  <= total_costs < 300000) Cst_Level = '09';
    when (300000  <= total_costs < 400000) Cst_Level = '10';
    otherwise cst_level = '99';
  end;

label Chg_Level = 'Charge Level'
      Cst_Level = 'Cost Level';
run;

* create permanent user defined formats for chg_level and cst_level ;
proc format library=library;
  value $chg '01' = 'Less than $1  '
             '02' = '    $1 to  $.5K'
             '03' = ' $.5K to   $1K'
             '04' = '  $1K to   $5K'
             '05' = '  $5K to  $10K'
             '06' = ' $10K to  $50K'
             '07' = ' $50K to $100K'
             '08' = '$100K to $200K'
             '09' = '$200K to $300K'
             '10' = '$300K to $400K'
             '99' = '        > $400K'
             ;
  value $cst '01' = 'Less than $1  '
```

```
                    '02' = '   $1 to   $.5K'
                    '03' = ' $.5K to    $1K'
                    '04' = '   $1K to    $5K'
                    '05' = '   $5K to   $10K'
                    '06' = '  $10K to   $50K'
                    '07' = '  $50K to  $100K'
                    '08' = '$100K to $200K'
                    '09' = '$200K to $300K'
                    '10' = '$300K to $400K'
                    '99' = '      > $400K'
                    ;
run;

title 'Frequency of Values for Total Charges and Total Cost';
proc freq data=HCDBD01A;
   tables chg_level
          cst_level;
   format chg_level $chg.
          cst_level $cst.;
run;
```

This program produces the report displayed below.

```
          Frequency of Values for Total Charges and Total Cost

                         The FREQ Procedure

                            Charge Level

                                             Cumulative   Cumulative
   Chg_Level           Frequency   Percent   Frequency     Percent

   Less than $1            60       0.57          60         0.57
      $1 to   $.5K         53       0.50         113         1.07
    $.5K to    $1K        383       3.63         496         4.70
      $1K to    $5K      2830      26.84        3326        31.54
      $5K to   $10K      2981      28.27        6307        59.82
     $10K to   $50K      3835      36.37       10142        96.19
     $50K to  $100K       310       2.94       10452        99.13
    $100K to  $200K        70       0.66       10522        99.79
    $200K to  $300K        11       0.10       10533        99.90
    $300K to  $400K         6       0.06       10539        99.95
          > $400K           5       0.05       10544       100.00


                             Cost Level

                                             Cumulative   Cumulative
   Cst_Level           Frequency   Percent   Frequency     Percent

   Less than $1             5       0.05           5         0.05
      $1 to   $.5K        272       2.58         277         2.63
    $.5K to    $1K        744       7.06        1021         9.68
      $1K to    $5K      5812      55.12        6833        64.80
      $5K to   $10K      2133      20.23        8966        85.03
     $10K to   $50K      1509      14.31       10475        99.35
     $50K to  $100K        53       0.50       10528        99.85
    $100K to  $200K        14       0.13       10542        99.98
    $200K to  $300K         1       0.01       10543        99.99
    $300K to  $400K         1       0.01       10544       100.00
```

From this report we see that low and high outliers occur for both variables. There are 496 discharges with charges less than $1,000 and 22 discharges with charges greater than of equal to $200,000. There are 1,021 discharges with costs less than $1,000 and 2 discharges with costs greater than of equal to $200,000.

Based on this information, a decision is made to exclude outliers from the subsequent calculations using the following outlier threshold definition.

Variable	Lower Threshold	Upper Threshold
Charges	< $1,000	$200,000

Manipulating the Data

HCDBM01A.sas

This program will read the SAS data set output by HCDBR01A.sas and keep only the discharges that fall within the outlier thresholds.

```
*-----------------------------------------------------------------------*
|                                                                       |
|Program:     HCDBM01A.sas                                              |
|                                                                       |
|Source:      c:\Projects\Health Care Cost\code\prod                    |
|                                                                       |
|Purpose:     Delete outliers.                                          |
|                                                                       |
|Input:       HCDBR01A.sas7bdat                                         |
|                                                                       |
|Output:      HCDBM01A.sas7bdat                                         |
|                                                                       |
|Update Log:  Notes                            Date        Programmer   |
|             ------------------------------   ----------  ------------ |
|             Original version                 06/04/2003  D. Blakeley  |
|                                                                       |
|Usage Notes: Discharges less than $1,000 are deleted.                  |
|             Discharges $200K or greater are deleted.                  |
|                                                                       |
|Other Notes:                                                           |
|                                                                       |
*-----------------------------------------------------------------------*;

libname sasin 'c:\Projects\Health Care Cost\data\prod';
run;

data sasin.HCDBM01A;
   * input discharges where charges are $1,000 or more but less than $200K ;
   set sasin.HCDBR01A (where=(1000 <= total_charges < 200000));
run;
```

The SAS log documents the results of outlier deletion. The output SAS data set containing only non-outlier discharges has 10,026 observations. 518 discharges have met the Charge and/or Cost outlier definitions.

```
NOTE: There were 10026 observations read from the data set SASIN.HCDBR01A.
      WHERE ((total_charges>=1000 and total_charges<200000));
NOTE: The data set SASIN.HCDBM01A has 10026 observations and 6 variables.
NOTE: DATA statement used:
      real time            0.11 seconds
      cpu time             0.04 seconds
```

Validating the Data

HCDBV02A.sas

The non-outlier data set will be validated using a program very similar to the first validation program (HCDBV01A.sas). The same set of descriptive statistics reports are produced by this program. Excerpts from the output are displayed below.

```
                     Summary of Total Charges and Costs by Year - HCDBM01A

        FILE_   DISCHARGE_
Obs     YEAR      MONTH      _FREQ_      Tot_Cgs         Tot_Cst        Ave_Cgs      Ave_Cst

 1      2000        12        5121      66814569.59    31587896.39     13047.17     6168.31
 2      2001        12        4905      70466179.81    30090948.51     14366.19     6134.75
```

The conclusions from the initial assessment are consistent with what we observe here.

```
                        The UNIVARIATE Procedure
                Variable:  TOTAL_CHARGES   (TOTAL_CHARGES)

                     Basic Statistical Measures

           Location                        Variability

   Mean      13692.47      Std Deviation                17279
   Median     8316.08      Variance                 298556632
   Mode       2072.90      Range                       195534
                           Interquartile Range          11219

                       Extreme Observations

      ------Lowest-----            -----Highest----

         Value      Obs            Value      Obs

        1002.25     9254          188261      4075
        1003.07     4717          190489      7583
        1003.25     5280          193087      4156
        1006.10     3231          194482      6766
        1006.80     5033          196536      2555
```

```
                        The UNIVARIATE Procedure
                Variable:  TOTAL_COSTS   (TOTAL_COSTS)

                     Basic Statistical Measures

           Location                        Variability

   Mean      6151.890      Std Deviation                 7724
   Median    3781.655      Variance                  59654851
   Mode       714.760      Range                       178294
                           Interquartile Range           4759

                       Extreme Observations

      ------Lowest-----            ------Highest-----

         Value      Obs            Value      Obs

       -2213.48     8389         91081.4      6766
       -1490.52     8418         92827.1      4167
        -372.37     9224         96446.3      6294
          90.09     5543        118323.9      4075
         101.66     6447        176080.7      4156
```

These reports indicate a decrease in variability for both Charges and Costs due to the elimination of low and high Charge outliers.

Other reports show the frequency of discharges by MDC and Census Region.

```
                           The FREQ Procedure

                                   MDC

                                          Cumulative    Cumulative
       MDC      Frequency      Percent    Frequency      Percent
```

MDC	Frequency	Percent	Cumulative Frequency	Cumulative Percent
1	542	5.41	542	5.41
10	266	2.65	808	8.06
11	323	3.22	1131	11.28
12	69	0.69	1200	11.97
13	338	3.37	1538	15.34
14	1249	12.46	2787	27.80
15	703	7.01	3490	34.81
16	92	0.92	3582	35.73
17	92	0.92	3674	36.64
18	195	1.94	3869	38.59
19	407	4.06	4276	42.65
2	14	0.14	4290	42.79
20	80	0.80	4370	43.59
21	132	1.32	4502	44.90

22	4	0.04	4506	44.94
23	145	1.45	4651	46.39
24	26	0.26	4677	46.65
25	27	0.27	4704	46.92
3	142	1.42	4846	48.33
4	1060	10.57	5906	58.91
5	1940	19.35	7846	78.26
6	828	8.26	8674	86.52
7	257	2.56	8931	89.08
8	883	8.81	9814	97.89
9	212	2.11	10026	100.00

CENSUS_REGION

CENSUS_REGION	Frequency	Percent	Cumulative Frequency	Cumulative Percent
1	253	2.52	253	2.52
2	3146	31.38	3399	33.90
3	5838	58.23	9237	92.13
4	789	7.87	10026	100.00

The elimination of outliers discharges has further reduced the number of discharges in several low volume MDCs.

Another perspective on the issues of variation and extreme values is represented by the plots of Total Charges (Figure AII.6) and Total Costs (Figure AII.7).

Figure AII.6

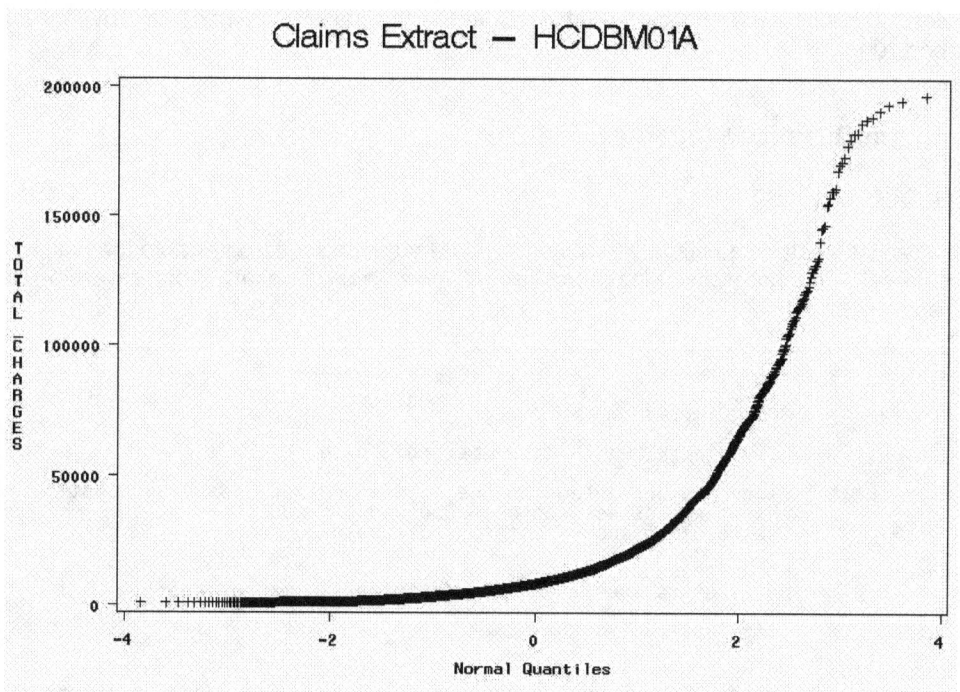

Figure AII.7

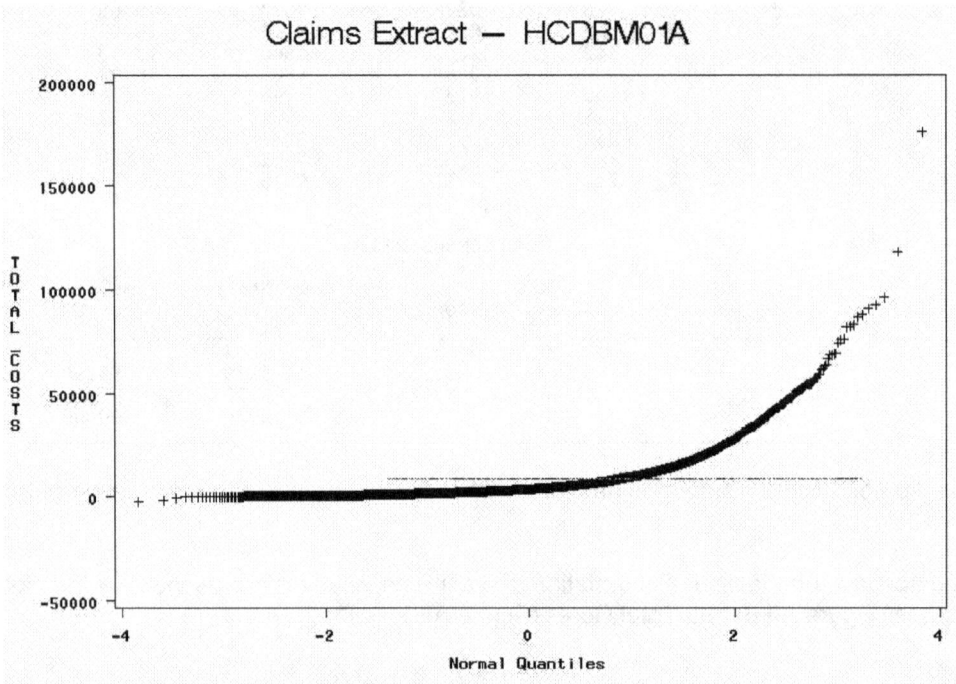

After a review of this new information with the requester, the decision is made to move forward with a second report run.

Answering the Questions (again)

HCDBP01B.sas

The macro "report" is modified slightly to accommodate a second macro variable that will represent the input data set. This program calls macro "report" twice to produce an updated version of the standard report set.

```
*---------------------------------------------------------------------*
Program:      HCDBP01B.sas

Source:       c:\Projects\Health Care Cost\code\prod

Purpose:      Plots the percentage change in mean charge, cost, and cost-
              charge ratios between Dec 2000 and Dec 2001.
              The level of analysis is the discharge.

Input:        HCDBP01B.mac
              Data set is specified as a parameter in macro report.

Output:       listings to output window

Update Log:   Notes                              Date          Programmer
              ------------------------------     ----------    ------------------
              Original version                   06/02/2003    D. Blakeley
              Added macro parameter for the      06/04/2003    D. Blakeley
              input data set name.

Usage Notes:  Specify the input data set name as the first parameter in
              macro report.
              Specify the level of analysis as the second parameter.

Other Notes:

*---------------------------------------------------------------------*;
options nodate mprint;
```

```
libname sasin 'c:\Projects\Health Care Cost\data\prod';
run;

%include 'c:\Projects\Health Care Cost\code\prod\HCDBP01B.mac';

%report(HCDBM01A,census_region)
%report(HCDBM01A,mdc)

quit;
```

A review of the report output from this program alleviates the concerns about excessive variability in the data set. However, upon further consideration, the requester concluded that the sample size for some MDCs was simply too low and requested that additional observations be added to the analysis data set. This change in scope was noted in the "Change History" section of the "Output Requirements" document.

The programmer was able to locate a data set from another project that contained 10,785 discharges corresponding to January 2001 and January 2002. This data set was derived from the same source as the December data using the same 3 percent sampling mechanism. The December 2000 data would be coupled with the January data and the December 2001 data would be coupled with the January 2002 data. However, before combining data sets, another set of validation statistics was run on the January data to assess it's compatibility with the December data.

HCDBV03A.sas

The standard set of validation statistics was produced for the January data. The program name was incremented from HCDBV02A.sas to HCBDV03A.sas to reflect that they have different input data sets. A review of the report output indicated that the same data issues were present in the January data as were observed in the December data. Therefore, the same outlier exclusion criteria will be applied when the two data sets are combined.

Combining the Data

HCDBC01A.sas

This program combines the December 2000/2001 and January 2001/2002 discharge data sets in order to increase overall sample size. Note that the December data has already been purged of outliers. Outliers in the January data will be deleted in this program.

```
*----------------------------------------------------------------------------*
  Program:      HCDBC01A.sas

  Source:       c:\Projects\Health Care Cost\code\prod

  Purpose:      Combine Dec. 00/01 discharge data with Jan. 01/02 discharge
                data to increase sample size.

  Input:        HCDBM01A.sas7bdat (Dec. 00/01 with outliers excluded)
                HCDBR01B.sas7bdat (Jan. 01/02 with outliers included)

  Output:       HCDBC01A.sas7bdat
                HCDBC01A_MDCFREQ.sas7bdat
                summary statistics to the output window

  Update Log:   Notes                              Date          Programmer
                ------------------------------     -----------   ------------------
                Original version                   08/05/2003    D. Blakeley

  Usage Notes:

  Other Notes:

*----------------------------------------------------------------------------*;

libname sasin 'c:\Projects\Health Care Cost\data\prod';
run;
```

```
* concatenate the Dec. and Jan. data sets ;
* exclude outliers from the Jan. data ;
data sasin.HCDBC01A;
  set sasin.HCDBM01A sasin.HCDBR01B (where=(1000 <= total_charges < 200000));
run;

* calculate frequency of discharges by MDC ;
proc summary data=sasin.HCDBC01A missing n nway;
  class mdc;
  var total_charges;
  output out=sasin.HCDBC01A_MDCFREQ (drop= _type_ _freq_ )n=freq;
run;

* print MDC frequency table ;
title 'MDC Frequency - HCDBC01A Data Set';
proc print data=sasin.HCDBC01A_MDCFREQ noobs;
run;
```

The SAS log documents the results of this combination.

```
NOTE: There were 10026 observations read from the data set SASIN.HCDBM01A.
NOTE: There were 10315 observations read from the data set SASIN.HCDBR01B.
      WHERE ((total_charges>=1000 and total_charges<200000));
NOTE: The data set SASIN.HCDBC01A has 20341 observations and 6 variables.
NOTE: DATA statement used:
      real time            0.31 seconds
      cpu time             0.06 seconds
```

The program outputs a permanent SAS data set containing the discharge frequency by MDC from the combined data set.

```
MDC Frequency - HCDBC01A Data Set

     MDC       freq

      1        1186
     10         600
     11         653
     12         145
     13         614
     14        2423
     15        1388
     16         200
     17         166
     18         386
     19         878
      2          24
     20         192
     21         273
     22          13
     23         250
     24          48
     25          50
      3         267
      4        2319
      5        3854
      6        1678
      7         522
      8        1833
      9         379
```

We can see from this listing that sample size has been augmented for the low volume MDCs, although not significantly.

Manipulating the Data (again)

HCDBM02A.sas

This program will manipulate the combined SAS data set to reformat the data for the purposes of final reporting. December 2000 discharges will be associated with the same reporting year as the January 2001 discharges. Likewise, the December 2001 discharges will be associated with the January 2002 reporting year. Any MDCs in the combined data set having less than 100 discharges will be flagged so that they stand out in the MDC reports.

```
*---------------------------------------------------------------------*
|                                                                     |
|Program:      HCDBM02A.sas                                           |
|                                                                     |
|Source:       c:\Projects\Health Care Cost\code\prod                 |
|                                                                     |
|Purpose:      Reformat variables for reporting purposes.             |
|                                                                     |
|Input:        HCDBC01A.sas7bdat                                      |
|              HCDBC01A_MDCFREQ.sas7bdat                              |
|                                                                     |
|Output:       HCDBM02A.sas7bdat                                      |
|              summary statistics to the output window                |
|                                                                     |
|Update Log:   Notes                          Date        Programmer  |
|              ----------------------------   ----------- ----------- |
|              Original version               08/05/2003  D. Blakeley |
|                                                                     |
|Usage Notes:                                                         |
|                                                                     |
|Other Notes:                                                         |
|                                                                     |
*---------------------------------------------------------------------*;

libname sasin 'c:\Projects\Health Care Cost\data\prod';
run;

proc sort data=sasin.HCDBC01A;
   by mdc;
run;

proc sort data=sasin.HCDBC01A_MDCFREQ;
   by mdc;
run;

* merge the MDC frequencies with the combined discharge data ;
data sasin.HCDBM02A;
   merge sasin.HCDBC01A sasin.HCDBC01A_MDCFREQ;
   by mdc;
   * flag any MDC's having less than 100 observations ;
   if freq < 100 then do;
     mdc = trim(mdc) || '*';
   end;
   * redefine file_year to combine (Dec. 00 + Jan. 01) and (Dec. 01 + Jan. 02) ;
   * this statement changes the Dec. 00 file year to 2001 ;
   if file_year = 2000 then file_year = 2001;
   * this statement changes the Dec. 01 file year to 2002 ;
   else
   if discharge_month = 12 then file_year = 2002;
run;

* calculate frequency of discharges by MDC ;
proc summary data=sasin.HCDBM02A missing n nway;
   class mdc;
   var total_charges;
   output out=HCDBM02A_MDCFREQ (drop= _type_ _freq_ )n=freq;
run;

* print MDC frequency table ;
title 'MDC Frequency with Low Volume Flags - HCDBM02A Data Set';
proc print data=HCDBM02A_MDCFREQ noobs;
run;

* calculate frequency of discharges by revised file_year ;
proc summary data=sasin.HCDBM02A missing n nway;
   class file_year;
   var total_charges;
   output out=HCDBM02A_YEARFREQ (drop= _type_ _freq_ )n=freq;
run;

* print file_year frequency table ;
title 'Revised file_year Frequency - HCDBM02A Data Set';
proc print data=HCDBM02A_YEARFREQ noobs;
run;
```

This program is an example of how data manipulation code intended to facilitate reporting can be more extensive than some of the serious "number crunching" programs. The output from this program shows the results of the data manipulation.

```
MDC Frequency with Low Volume Flags - HCDBM02A Data Set
                    MDC      freq

                    1        1186
                    10        600
                    11        653
                    12        145
                    13        614
                    14       2423
                    15       1388
                    16        200
                    17        166
                    18        386
                    19        878
                    2*         24
                    20        192
                    21        273
                    22*        13
                    23        250
                    24*        48
                    25*        50
                    3         267
                    4        2319
                    5        3854
                    6        1678
                    7         522
                    8        1833
                    9         379


        Revised file_year Frequency - HCDBM02A Data Set

                    FILE_
                    YEAR      freq

                    2001     10886
                    2002      9455
```

Four low volume MDCs are flagged and the counts of the reconstituted pairs of December and January discharges are displayed. How do we know if those counts are correct? Is there a way to reconcile these totals to ensure that we are correctly accounting for the discharges associated with each period? The next program will tackle this task.

Validating the Data (again)

HCDBV04A.sas

This program accounts for all deletions and data transformations in the combined discharge data set. We start by concatenating the original December and January discharge data sets, which include outliers. Each discharge is categorized as a "Non-Outlier" or "Outlier".

```
*-----------------------------------------------------------------------*
|                                                                       |
| Program:     HCDBV04A.sas                                             |
|                                                                       |
| Source:      c:\Projects\Health Care Cost\code\prod                   |
|                                                                       |
| Purpose:     Accounts for all deletions and data transformations in the |
|              combined data set.                                       |
|                                                                       |
| Input:       HCDBR01A.sas7bdat (complete Dec. 00 and Dec. 01 discharge file) |
|              HCDBR01B.sas7bdat (complete Jan. 01 and Jan. 02 discharge file) |
|                                                                       |
| Output:      Summary statistics to the output window                  |
|                                                                       |
*-----------------------------------------------------------------------*
```

```
|Update Log:  Notes                              Date        Programmer        |
|             -----------------------------      ----------  ----------------  |
|             Original version                   08/05/2003  D. Blakeley       |
|                                                                              |
|Usage Notes:                                                                  |
|                                                                              |
|Other Notes:                                                                  |
|                                                                              |
*------------------------------------------------------------------------------*;
libname sasin 'c:\Projects\Health Care Cost\data\prod';
run;

data check;
  * delete outliers ;
  set sasin.HCDBR01A
      sasin.HCDBR01B;
  length category $ 11;
  * flag outliers ;
  if (1000 <= total_charges < 200000) then category = 'Non-Outlier';
  else category = 'Outlier';
run;

proc summary data=check missing n;
  class category discharge_month file_year;
  var total_charges;
  output out=stats (drop= _type_ _freq_ ) n=freq;
run;

title 'Summary of Total Discharges';
proc print data=stats;
run;
```

The data set is then summarized to count the frequency of all possible combinations of outlier category, discharge month, and year. The output is displayed below.

```
                    Summary of Total Discharges

                           DISCHARGE_   FILE_
   Obs    category          MONTH       YEAR      freq

     1                         .          .       21329
     2                         .        2000       5408
     3                         .        2001      11192
     4                         .        2002       4729
     5                         1          .       10785
     6                        12          .       10544
     7                         1        2001       6056
     8                         1        2002       4729
     9                        12        2000       5408
    10                        12        2001       5136
    11    Non-Outlier          .          .       20341
    12    Outlier              .          .         988
    13    Non-Outlier          .        2000       5121
    14    Non-Outlier          .        2001      10670
    15    Non-Outlier          .        2002       4550
    16    Outlier              .        2000        287
    17    Outlier              .        2001        522
    18    Outlier              .        2002        179
    19    Non-Outlier          1          .       10315
    20    Non-Outlier         12          .       10026
    21    Outlier              1          .         470
    22    Outlier             12          .         518
    23    Non-Outlier          1        2001       5765
    24    Non-Outlier          1        2002       4550
    25    Non-Outlier         12        2000       5121
    26    Non-Outlier         12        2001       4905
    27    Outlier              1        2001        291
    28    Outlier              1        2002        179
    29    Outlier             12        2000        287
    30    Outlier             12        2001        231
```

It's now a fairly simple matter to reconcile these discharge counts to ensure that we have processed the data correctly.

Period	Total Discharges	Non-Outliers
Dec 2000	5408	5121
Jan 2001	6056	5765
Dec 2001	5136	4905
Jan 2002	4729	4550
	21329	20341

The Non-Outlier total of 20,341 discharges matches the number of observations in the HCDBM02A data set. We now have an "Analysis Dataset" with all the elements necessary to deliver the reports as specified.

Answering the Questions (yet again)

Reporting

RUDBP01C.sas

This is the third and final version of the reporting program. As you can see from the update log in the program header, several modifications have occurred to make the processing more data driven.

```
*-----------------------------------------------------------------------*
|                                                                       |
| Program:      HCDBP01C.sas                                            |
|                                                                       |
| Source:       c:\Projects\Health Care Cost\code\prod                  |
|                                                                       |
| Purpose:      Plots the percentage change in mean charge, cost, and cost- |
|               charge ratios between Dec 2000 and Dec 2001.            |
|               The level of analysis is the discharge.                 |
|                                                                       |
| Input:        HCDBP01C.mac                                            |
|               Data set is specified as a parameter in macro report.   |
|                                                                       |
| Output:       listings to output window                               |
|                                                                       |
| Update Log:   Notes                         Date         Programmer   |
|               -----------------------------  -----------  ------------ |
|               Original version              08/04/2003   D. Blakeley  |
|               Added macro parameter for the 08/06/2003   D. Blakeley  |
|               input data set name.                                    |
|               Added macro parameters for the 08/07/203   D. Blakeley  |
|               'from' and 'to' period.                                 |
|               Added a footnote.                                       |
|                                                                       |
| Usage Notes:  Specify the input data set name as the first parameter in |
|               macro report.                                           |
|               Specify the level of analysis as the second parameter.  |
|               Specify the from and to date periods as the third and fourth |
|               parameters.  Use the format MMMYY-MMMYY.                |
|                                                                       |
| Other Notes:                                                          |
|                                                                       |
*-----------------------------------------------------------------------*;
options nodate mprint;
footnote;

libname sasin 'c:\Projects\Health Care Cost\data\prod';
run;

%include 'c:\Projects\Health Care Cost\code\prod\HCDBP01C.mac';

%report(HCDBM02A,census_region,Dec00-Jan01,Dec01-Jan02)
footnote 'Low Volume MDCs Are Identified by an Asterisk';
%report(HCDBM02A,mdc,Dec00-Jan01,Dec01-Jan02)

quit;
```

As with the earlier versions of this program and the associated macro, a set of listings and plots are output for Census Region and MDC. The MDC reports in their entirety are displayed on the following pages.

```
                              Health Care Cost Project
                              Input Data Set - HCDBM02A
                       Data for Dec00-Jan01 and Dec01-Jan02
                         Percentage Changes by mdc and Year
```

MDC	Claim Year	Patient Discharges	Average Charge	Average Cost	Cost-to-Charge Ratio	% Change Charges	% Change Costs	% Change C-C Ratio
1	2001	678	15257.57	6815.38	0.44669	.	.	.
1	2002	508	15911.85	6577.75	0.41339	4.29%	(3.49%)	(7.46%)
10	2001	319	9950.13	4753.73	0.47776	.	.	.
10	2002	281	12312.18	4961.28	0.40296	23.74%	4.37%	(15.66%)
11	2001	330	13619.57	5831.69	0.42818	.	.	.
11	2002	323	13965.21	5925.88	0.42433	2.54%	1.62%	(0.90%)
12	2001	86	11445.21	5393.44	0.47124	.	.	.
12	2002	59	10914.08	4418.57	0.40485	(4.64%)	(18.08%)	(14.09%)
13	2001	311	9972.04	4260.35	0.42723	.	.	.
13	2002	303	10194.05	4210.77	0.41306	2.23%	(1.16%)	(3.32%)
14	2001	1323	5680.99	3079.67	0.54210	.	.	.
14	2002	1100	6485.10	3328.17	0.51320	14.15%	8.07%	(5.33%)
15	2001	713	5992.26	3132.71	0.52279	.	.	.
15	2002	675	5329.34	2580.28	0.48416	(11.06%)	(17.63%)	(7.39%)
16	2001	100	12527.00	5494.46	0.43861	.	.	.
16	2002	100	11700.11	4465.32	0.38165	(6.60%)	(18.73%)	(12.99%)
17	2001	96	18648.22	8078.44	0.43320	.	.	.
17	2002	70	23467.33	9868.62	0.42053	25.84%	22.16%	(2.93%)
18	2001	207	16364.60	7341.44	0.44862	.	.	.
18	2002	179	20561.85	8022.23	0.39015	25.65%	9.27%	(13.03%)
19	2001	442	7913.30	4715.53	0.59590	.	.	.
19	2002	436	7688.92	4352.79	0.56611	(2.84%)	(7.69%)	(5.00%)
2*	2001	10	9804.79	4704.83	0.47985	.	.	.
2*	2002	14	11522.93	4634.19	0.40217	17.52%	(1.50%)	(16.19%)
20	2001	112	5439.85	2898.16	0.53276	.	.	.
20	2002	80	5892.71	3297.78	0.55964	8.32%	13.79%	5.04%
21	2001	133	11102.84	5195.68	0.46796	.	.	.
21	2002	140	12658.55	5459.97	0.43133	14.01%	5.09%	(7.83%)
22*	2001	9	5413.84	3289.52	0.60761	.	.	.
22*	2002	4	28259.04	14019.66	0.49611	421.98%	326.19%	(18.35%)
23	2001	125	15874.66	8929.37	0.56249	.	.	.
23	2002	125	16645.38	8841.28	0.53116	4.86%	(0.99%)	(5.57%)
24*	2001	26	38572.61	15593.30	0.40426	.	.	.
24*	2002	22	38449.91	16384.21	0.42612	(0.32%)	5.07%	5.41%
25*	2001	29	19342.39	7912.48	0.40907	.	.	.
25*	2002	21	24481.23	8754.81	0.35761	26.57%	10.65%	(12.58%)
3	2001	157	11688.56	5490.81	0.46976	.	.	.
3	2002	110	11827.89	4471.55	0.37805	1.19%	(18.56%)	(19.52%)
4	2001	1195	13708.93	6041.80	0.44072	.	.	.
4	2002	1124	14553.39	5922.30	0.40694	6.16%	(1.98%)	(7.67%)
5	2001	2101	19846.72	8613.60	0.43401	.	.	.
5	2002	1753	21406.56	8517.69	0.39790	7.86%	(1.11%)	(8.32%)
6	2001	876	14432.19	6506.33	0.45082	.	.	.
6	2002	802	15354.88	6122.07	0.39870	6.39%	(5.91%)	(11.56%)
7	2001	273	16831.55	7041.79	0.41837	.	.	.
7	2002	249	16527.32	6629.68	0.40113	(1.81%)	(5.85%)	(4.12%)
8	2001	1009	17066.13	7637.05	0.44750	.	.	.
8	2002	824	20068.39	8495.96	0.42335	17.59%	11.25%	(5.40%)
9	2001	226	11288.78	5175.25	0.45844	.	.	.
9	2002	153	11749.12	4843.10	0.41221	4.08%	(6.42%)	(10.08%)

```
                   Low Volume MDCs Are Identified by an Asterisk
```

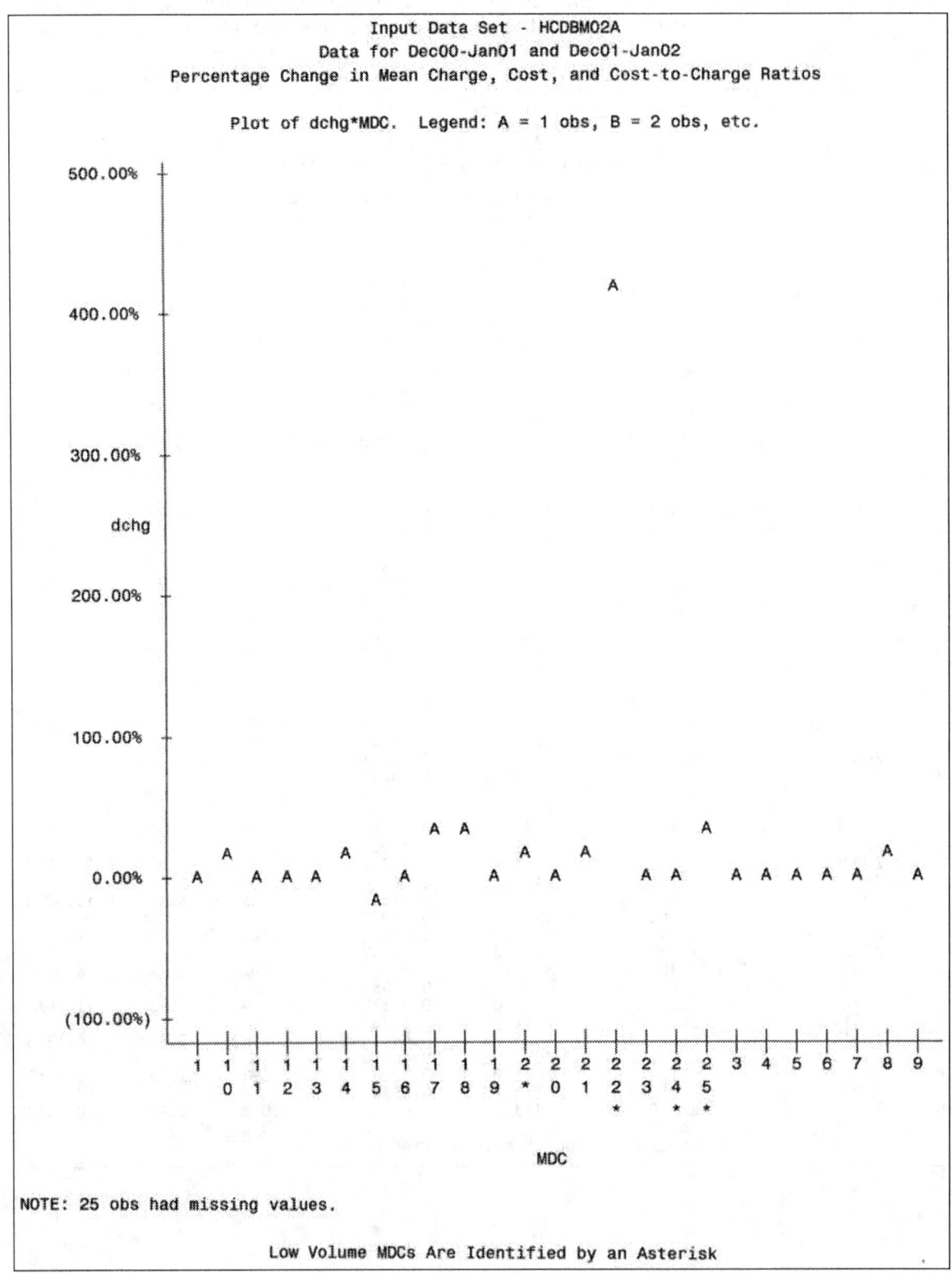

Input Data Set - HCDBM02A
Data for Dec00-Jan01 and Dec01-Jan02
Percentage Change in Mean Charge, Cost, and Cost-to-Charge Ratios

Plot of dchg*MDC. Legend: A = 1 obs, B = 2 obs, etc.

NOTE: 25 obs had missing values.

Low Volume MDCs Are Identified by an Asterisk

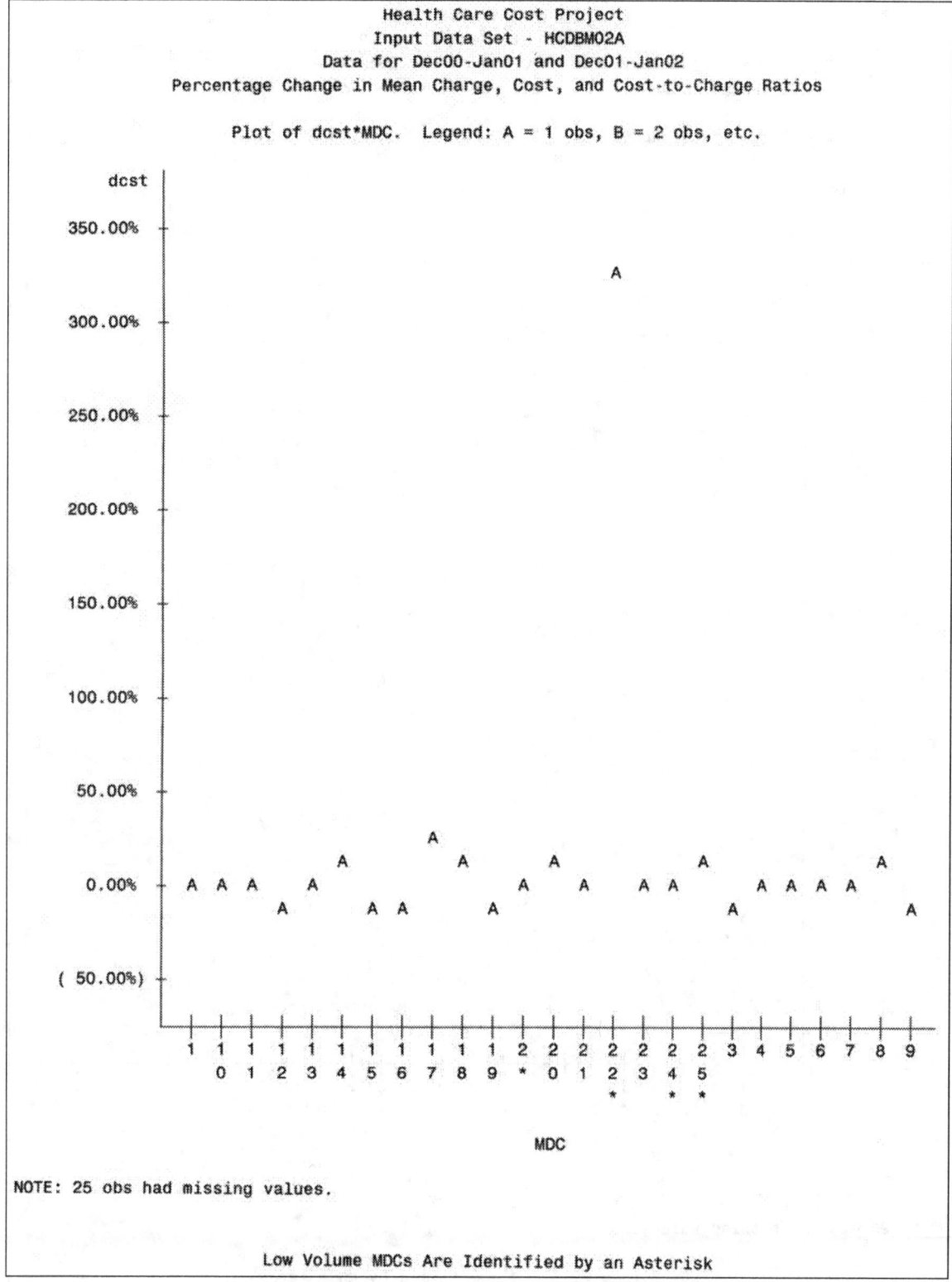

Health Care Cost Project
Input Data Set - HCDBM02A
Data for Dec00-Jan01 and Dec01-Jan02
Percentage Change in Mean Charge, Cost, and Cost-to-Charge Ratios

Plot of dcst*MDC.   Legend: A = 1 obs, B = 2 obs, etc.

NOTE: 25 obs had missing values.

Low Volume MDCs Are Identified by an Asterisk

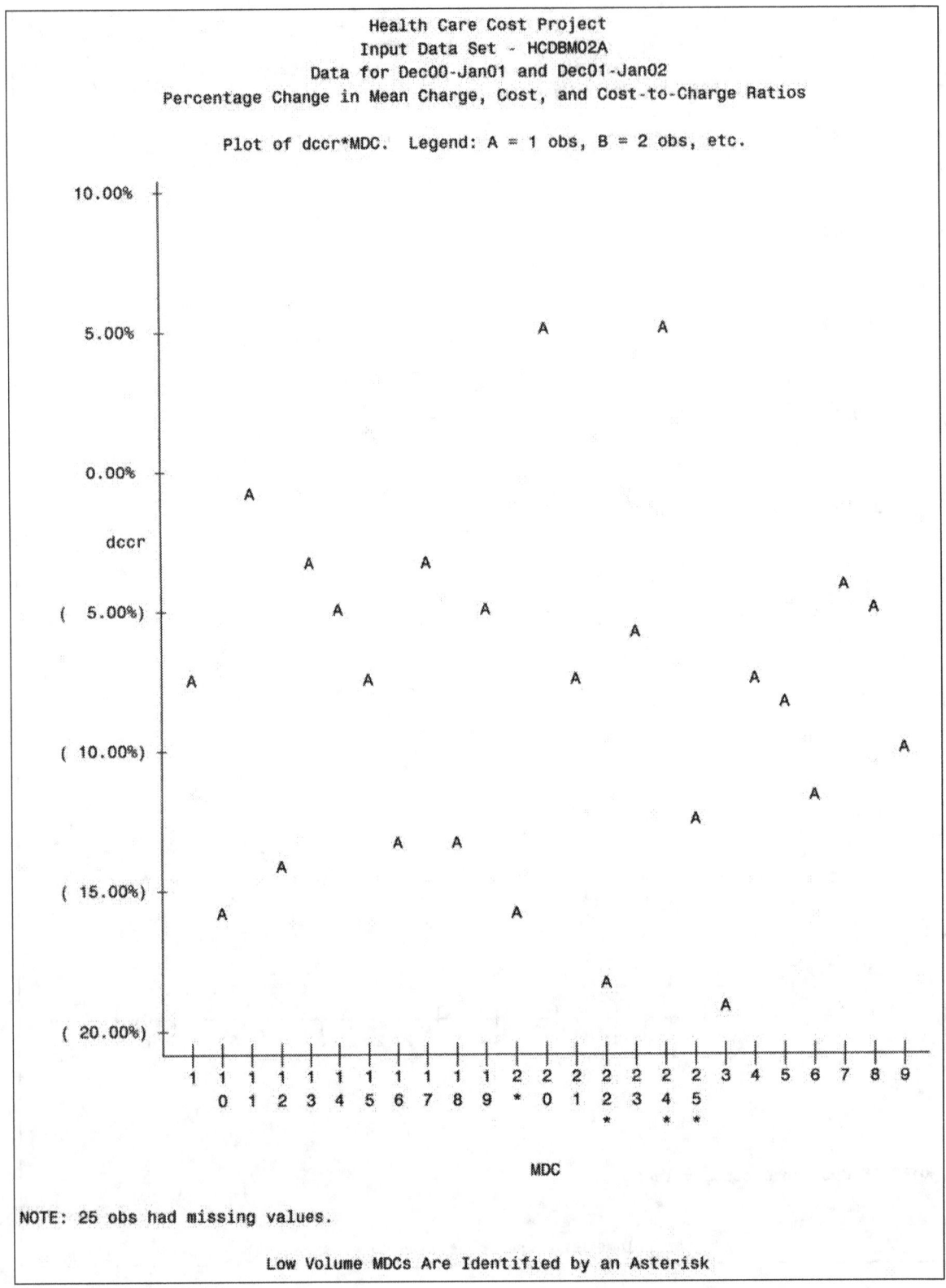

What do these results tell us? In general we observe that average charges are increasing for most MDCs, many at rates greater than 25 percent. Average costs are also increasing for most MDCs, but generally at a slower rate than charges. This comparison of the rate of increase in cost versus charges is further confirmed by the observed change in cost-to-charge ratios. We see that most cost-to-charge rations are decreasing, which occurs when charges rise faster than costs. The change in cost-to-charge ratios is positive when costs are rising faster than charges. It's also important to note that the extreme results for MDC 22 should be interpreted with caution. In spite of tripling the sample size for this MDC, the number of discharges is still quite small making the resulting statistics potentially unreliable indicators.

Staying Organized

Building the Audit Trail

Using a program run log was useful in this project, particularly since there was a significant time gap between the start and end of the project due to the need to increase sample size. Listed below is the chronological sequence of events that occurred in the course of this project. The log distinguishes test runs from production runs, includes comments that are helpful in reconstructing events, and otherwise captures information that can come in handy when sorting through output and other documentation. For example, there were three iterations of the HCDBP01* program series. Only the final iteration, using the HCDBP01C version is relevant to the final results. Therefore, any output associated with the earlier versions can be discarded from final project documentation.

Date	Program	Type	Comments	Results
5/29/03	HCDBT01A	T	Assumed file was 1999-2001	Data from 10/00-6/02 (21 mos.)
5/29/03	HCDBR01A	P	Dec. 2000/01 discharges	10544 obs.
5/29/03	HCDBV01A	P	Initial validation	Looks suspicious
6/02/03	HCDBP01A	P	1st report run	Extreme obs and variability. Requester has questions.
6/04/03	HCDBD01A	P	Outlier analysis	O.K.
6/04/03	HCDBM01A	P	Exclude outliers	O.K.
6/04/03	HCDBV02A	P	Re-run validation	O.K.
6/04/03	HCDBP01B	P	2nd report run (with new macro variable)	O.K.
8/04/03	HCDBV03A	P	Jan 2001/02 discharges	10785 obs.
8/05/03	HCDBC01A	P	Concat Dec & Jan	20811 obs.
8/05/03	HCDBM02A	P	Reformat for reporting	O.K.
8/05/03	HCDBV04A	P	Final record count check	O.K.
8/07/03	HCDBP01C	P	Final report run	O.K.

You can see from the file dates in the project folders below that some of the programs were subsequently rerun.

Other steps taken during the course of the project include:
- Saving interim steps in the form of data sets or report listings. See the flowchart below.
- Updating the documentation block for each program to accurately reflect inputs, outputs, and changes.
- Ensuring that the programs are adequately commented.
- Saving test results that document the accuracy of your programming.

Managing Files

The folder configuration displayed in Figure AII.2 was used to organize and file all project-specific code, data, documentation, and formats. Folder contents at the conclusion of the project are displayed below.

Projects\Health Care Cost\code\prod

Name	Size	Type	Modified
HCDBC01A.sas	3 KB	SAS File	8/8/03 10:02 AM
HCDBD01A.sas	5 KB	SAS File	8/7/03 9:01 AM
HCDBM01A.sas	2 KB	SAS File	8/7/03 9:09 AM
HCDBM02A.sas	4 KB	SAS File	8/8/03 10:03 AM
HCDBP01A.mac	4 KB	MAC File	8/7/03 8:42 AM
HCDBP01A.sas	3 KB	SAS File	8/7/03 8:43 AM
HCDBP01B.mac	4 KB	MAC File	8/7/03 9:21 AM
HCDBP01B.sas	3 KB	SAS File	6/4/03 4:58 PM
HCDBP01C.mac	4 KB	MAC File	8/8/03 10:06 AM
HCDBP01C.sas	3 KB	SAS File	8/8/03 10:05 AM
HCDBR01A.sas	3 KB	SAS File	8/8/03 9:56 AM
HCDBV01A.sas	3 KB	SAS File	8/8/03 9:57 AM
HCDBV02A.sas	3 KB	SAS File	6/4/03 5:06 PM
HCDBV03A.sas	3 KB	SAS File	8/8/03 9:58 AM
HCDBV04A.sas	3 KB	SAS File	8/8/03 10:04 AM

Projects\Health Care Cost\code\test

Name	Size	Type	Modified
HCDBT01A.sas	3 KB	SAS File	8/7/03 8:38 AM
Program Doc Template.txt	2 KB	Text Document	5/13/03 2:49 PM

Projects\Health Care Cost5\data\prod

Name	Size	Type	Modified
hcdbc01a.sas7bdat	809 KB	SAS7BDAT File	8/7/03 9:41 AM
hcdbc01a_mdcfreq.sas7bdat	5 KB	SAS7BDAT File	8/7/03 9:41 AM
hcdbm01a.sas7bdat	401 KB	SAS7BDAT File	8/7/03 9:10 AM
hcdbm02a.sas7bdat	973 KB	SAS7BDAT File	8/7/03 9:41 AM
hcdbr01a.sas7bdat	421 KB	SAS7BDAT File	5/29/03 5:58 PM
hcdbr01b.sas7bdat	433 KB	SAS7BDAT File	5/29/03 5:39 PM

Projects\Health Care Cost\data\test -- No files.

Projects\Health Care Cost\doc

Name	Size	Type	Modified
HCDBC01A.log	4 KB	LOG File	8/7/03 9:31 AM
HCDBC01A.lst	2 KB	LST File	8/7/03 9:31 AM
HCDBD01A.log	6 KB	LOG File	8/7/03 9:03 AM
HCDBD01A.lst	3 KB	LST File	8/7/03 9:03 AM
HCDBM01A.log	3 KB	LOG File	8/7/03 9:11 AM
HCDBM02A.log	6 KB	LOG File	8/7/03 9:42 AM
HCDBM02A.lst	2 KB	LST File	8/7/03 9:42 AM
HCDBP01A.log	9 KB	LOG File	8/7/03 8:54 AM
HCDBP01A.lst	17 KB	LST File	8/7/03 8:55 AM
HCDBP01B.log	10 KB	LOG File	8/7/03 9:24 AM
HCDBP01B.lst	17 KB	LST File	8/7/03 9:24 AM
HCDBP01C.log	10 KB	LOG File	8/7/03 9:51 AM
HCDBP01C.lst	18 KB	LST File	8/7/03 9:51 AM
HCDBR01A.log	4 KB	LOG File	5/29/03 6:01 PM
HCDBR01A.lst	2 KB	LST File	5/29/03 6:01 PM
HCDBR01B.log	4 KB	LOG File	5/29/03 5:42 PM
HCDBR01B.lst	2 KB	LST File	5/29/03 5:42 PM
HCDBT01A.log	3 KB	LOG File	6/3/03 3:34 PM
HCDBT01A.lst	17 KB	LST File	6/3/03 3:34 PM
HCDBV01A.log	4 KB	LOG File	6/4/03 3:47 PM
HCDBV01A.lst	10 KB	LST File	6/4/03 3:47 PM
HCDBV01A_G1a.bmp	1,255 KB	Bitmap Image	6/4/03 3:49 PM
HCDBV01A_G1b.bmp	1,255 KB	Bitmap Image	6/4/03 3:50 PM
HCDBV02A.log	4 KB	LOG File	8/7/03 9:17 AM
HCDBV02A.lst	10 KB	LST File	8/7/03 9:17 AM
HCDBV02A_G2a.bmp	1,255 KB	Bitmap Image	8/7/03 9:18 AM
HCDBV02A_G2b.bmp	1,255 KB	Bitmap Image	8/7/03 9:19 AM
HCDBV03A.log	4 KB	LOG File	8/7/03 4:00 PM
HCDBV03A.lst	11 KB	LST File	8/7/03 4:00 PM
HCDBV03A_G3a.bmp	1,255 KB	Bitmap Image	8/7/03 4:01 PM
HCDBV03A_G3b.bmp	1,255 KB	Bitmap Image	8/7/03 4:01 PM
HCDBV04A.log	4 KB	LOG File	8/7/03 9:47 AM
HCDBV04A.lst	3 KB	LST File	8/7/03 9:48 AM
Health Care Cost Flow.vsd	66 KB	Microsoft Visio Drawi...	8/8/03 10:44 AM
PROD_DATA_CONTENTS.lst	14 KB	LST File	8/8/03 11:27 AM

138

Projects\Health Care Cost\formats

Name ▲	Size	Type	Modified
formats.sas7bcat	17 KB	SAS7BCAT File	8/7/03 9:02 AM

Managing Change

The primary changes in scope encountered in this project were the decisions to delete outliers and increase sample size. Both factors increased the amount of programming and iteration before the Analysis Data Set was finalized. Note that this change, which included a significant revision to the project due date, was recorded in the "Output Requirements" form displayed in Figure AII.1.

Documenting the Results

Making the Results Replicable

A review of the folder contents above shows that for each program we have saved the following information:
- Final code (in the code\prod folder)
- SAS Logs (in the doc folder)
- Sample prints (in the doc folder)

The CONTENTS and DATASETS Procedures

We have also used the CONTENTS procedure to document each production SAS data set using the code displayed below.

```
libname sasin 'c:\Projects\Health Care Cost\data\prod';
proc contents data=sasin._all_;
run;
```

Partial output from the CONTENTS procedure is displayed below

Directory Listing of the data\prod Folder

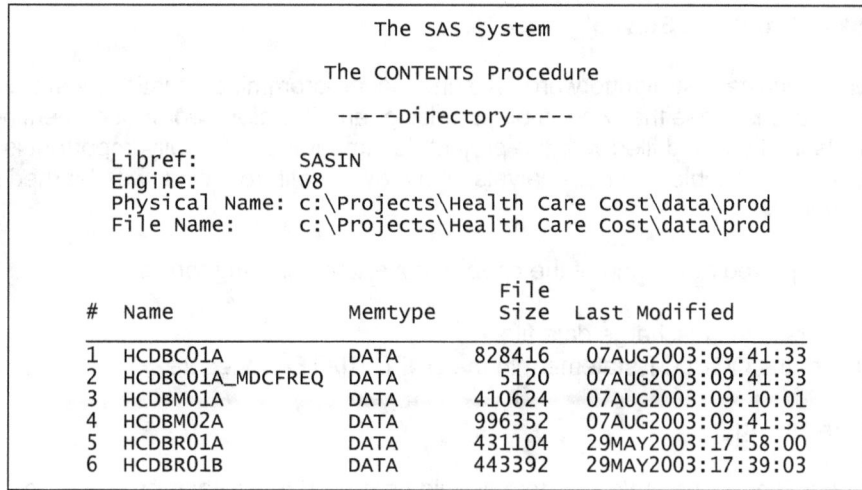

```
                        The SAS System

                      The CONTENTS Procedure

                       -----Directory-----

          Libref:         SASIN
          Engine:         V8
          Physical Name:  c:\Projects\Health Care Cost\data\prod
          File Name:      c:\Projects\Health Care Cost\data\prod


                                        File
        #  Name               Memtype   Size   Last Modified
        ───────────────────────────────────────────────────────
        1  HCDBC01A           DATA      828416  07AUG2003:09:41:33
        2  HCDBC01A_MDCFREQ   DATA        5120  07AUG2003:09:41:33
        3  HCDBM01A           DATA      410624  07AUG2003:09:10:01
        4  HCDBM02A           DATA      996352  07AUG2003:09:41:33
        5  HCDBR01A           DATA      431104  29MAY2003:17:58:00
        6  HCDBR01B           DATA      443392  29MAY2003:17:39:03
```

Contents of the "Analysis Dataset" HCDBM02A.sas7bdat

```
                      The CONTENTS Procedure

     Data Set Name: SASIN.HCDBM02A              Observations:          20341
     Member Type:   DATA                        Variables:             7
     Engine:        V8                          Indexes:               0
     Created:       9:41 Thursday, August 7, 2003   Observation Length:    48
     Last Modified: 9:41 Thursday, August 7, 2003   Deleted Observations: 0
     Protection:                                Compressed:            NO
     Data Set Type:                             Sorted:                NO
     Label:


                -----Engine/Host Dependent Information-----

     Data Set Page Size:          4096
     Number of Data Set Pages:    243
     First Data Page:             1
     Max Obs per Page:            84
     Obs in First Data Page:      46
     Number of Data Set Repairs:  0
     File Name:                   c:\Projects\Health Care Cost\data\prod\hcdbm02a.sas7bdat
     Release Created:             8.0202M0
     Host Created:                WIN_PRO


                -----Alphabetic List of Variables and Attributes-----

     #   Variable          Type   Len   Pos   Format   Informat   Label
    ─────────────────────────────────────────────────────────────────────────
     6   CENSUS_REGION     Char    2    44    $2.      $2.        CENSUS_REGION
     2   DISCHARGE_MONTH   Num     8     8    4.       4.         DISCHARGE_MONTH
     1   FILE_YEAR         Num     8     0    5.       5.         FILE_YEAR
     3   MDC               Char    4    40    $4.      $4.        MDC
     4   TOTAL_CHARGES     Num     8    16    14.2     14.2       TOTAL_CHARGES
     5   TOTAL_COSTS       Num     8    24    14.2     14.2       TOTAL_COSTS
     7   freq              Num     8    32                        TOTAL_CHARGES
```

The CONTENTS procedure output should also be saved in the doc folder.

Flowcharting

The project flowchart was prepared to document each input, program, and output. It is displayed in Figure AII.8.

Keeping Track of Problems Solved

This project was technically straightforward. No unusual programming situations were encountered. However, two situations arose that were a bit surprising, and therefore worth documenting. Both are captured in an "Issue Log" and filed with the project documentation. Equally important is to make issue log information available to other analysts who may benefit from what was learned in the course of this project.

The Issue Logs displayed at the end of the chapter cover the following topics:

- Outliers in the discharge data file.
- Use of the QQPLOT statement in the UNIVARIATE procedure.

Final Project Archiving

All of the files contained in the project folders should be copied for storage on a separate device and/or physical location.

Figure AII.8

Health Care Cost

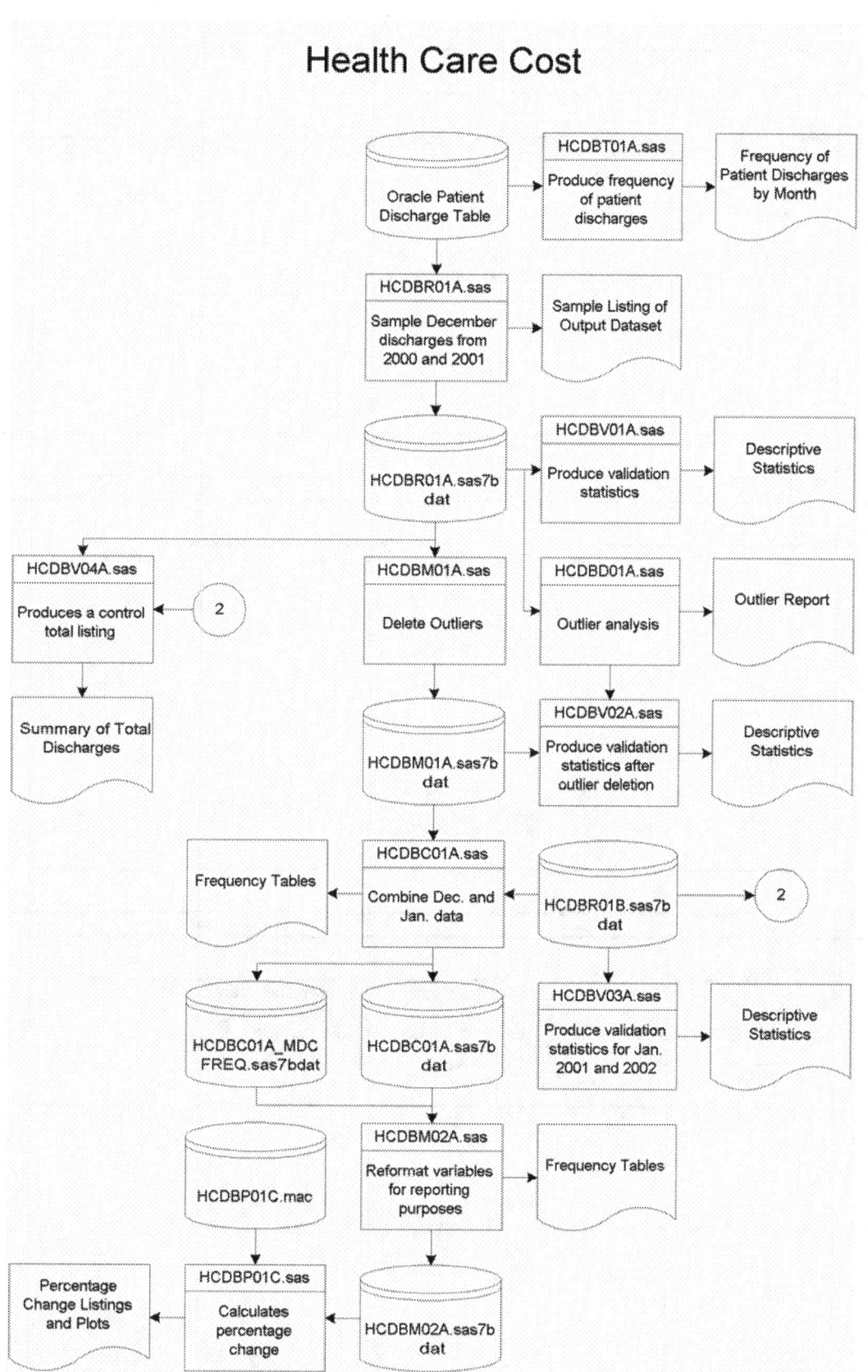

Issue Log - Hospital Discharge Charge Outliers

Date: 8/8/03	Project: Health Care Cost
Initiator Name	D. Blakeley

1. Description:

High and low outliers were discovered in the December 2000 and December 2001 discharge data extracted from the National Hospital Claims files. The findings are as follows:

- 4.7% of discharges had Total Charges less than $1000
- .2% of discharges had Total Charges greater than or equal to $200000
- 9.68% of discharges had Total Costs less than $1000
- .02% of discharges had Total Costs greater than or equal to $200000

2. Turnover:

Issue Passed To: File	Date: 8/8/03

3. Resolution:

Outlier cases were deleted if Total Charges were less than $1000 or greater than or equal to $200000.

4. Return:

Issue Received From: n/a	Date: n/a

5. Classification - check all that apply:

Data	☐ Missing Values	☐ Wrong Dataset	☐ Incomplete Data
	☐ Different Format	☐ Unknown Values	☒ Other - outliers
Coding	☐ Faulty Logic	☐ Used Wrong Data	☐ Formula Error
	☐ Selection Error	☐ Other	
Infrastructure	☐ Memory	☐ Space	☐ Other
Other			

Issue Log - QQPLOT Statement in the UNIVARIATE Procedure

Date: 8/8/03	Project: Health Care Cost
Initiator Name	D. Blakeley

1. Description:

The qqplot statement in the UNIVARIATE procedure creates a quantile-quantile plot that compares ordered variable values with quantiles of a specified theoretical distribution. The Q-Q plots are high-resolution graphics that do not appear in the SAS Output window with other procedure output. Instead, they appear in the SAS Graph window and must be saved as separate files.

2. Turnover:

Issue Passed To: File	Date: 8/8/03

3. Resolution:

When saving for use in project documentation, select the "Export as Image" option in the "File" menu.

4. Return:

Issue Received From: n/a	Date: n/a

5. Classification - check all that apply:

Data	☐ Missing Values ☐ Wrong Dataset ☐ Incomplete Data ☐ Different Format ☐ Unknown Values ☐ Other
Coding	☐ Faulty Logic ☐ Used Wrong Data ☐ Formula Error ☐ Selection Error ☒ Other - PROC UNIVARIATE QQPLOT statement
Infrastructure	☐ Memory ☐ Space ☐ Other
Other	

Case Study III

Introduction

This case study uses an extract from a Lotus Notes® application to produce reports which quantify Information Technology staff time and expense. The level of analysis is the individual project. The data are aggregated into two views: the semi-monthly payroll period and by calendar month.

Defining the Questions

The requester of these reports has a single primary question to be answered: "What are the labor hours and estimated cost incurred for each project calculated on a per pay period basis and monthly basis?" Use the "Output Requirements" template (Figure AIII.1) to capture this and other related information.

Getting Organized

Program Names

A Project Code of "RU" will be used as the first two characters in each program name. Since the programmer is D. Blakeley, the third and fourth characters in each program name will be "DB". The combination of Task Type, Sequence Number, and Version Number is specific to each program. In all cases the extension for SAS programs will be "sas". Therefore, the first program for this project will be "RUDBT01A.sas", defined as:

RUDBT01A =	RU	= Resource Utilization project
	DB	= D. Blakeley
	T	= test reading data
	01	= first program in the sequence
	A	= version A of the program

Output Files

Outputs will be prefaced with the characters of the program from which they originate. For example, the output data set from the second program ("RUDBR01A.sas") will be called "RUDBR01A" and will be written as RUDBR01A.sas7bdat.

Creating Folders

The standard set of folders, plus the addition of a folder for user-defined formats was created for this project (Figure AIII.2).

Figure AIII.2

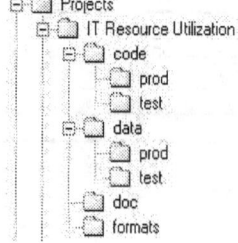

145

Figure AIII.1

Output Requirements

Date: 5/1/2003	Requester: R. Powers	Analyst: D. Blakeley	
Project: IT Resource Utilization	Report Name: Project Totals		Rpt. # 001
Business Purpose	Quantify IT resource utilization by project.		
Question(s) to be Answered	What are the labor hours and estimated cost incurred for each project calculated on a per pay period basis and monthly basis?		

1. Definitions:

Inputs	1. Time Tracking system (Lotus Notes application) extract 2. Hourly rates for contractors and daily rates for employees 3. Project names and types
Assumptions	1. All personnel have consistently and accurately entered their time during the analysis period of 6/15/2002-11/15/2002. 2. Contractor rates for the analysis period are valid.
Definitions	1. Projects with ptype = 'B' are "Business Change Projects" 2. Projects with ptype = 'S' are "Support Projects" 3. Contractors are identified as rtype = 'C' 4. Employees are identified as rtype = 'E'
Calculations	For periods before 1/1/2003: Contractor cost = hours multiplied by 1996 hourly rate Employee cost = convert the daily rate to an hourly equivalent by multiplying it by a factor of 1000/2080, then multiplying the result by the benefits loading factor of 1.32, then multiply the result by project hours. For periods after 12/31/2002 (if any): Same as above except use 2003 rates (if available).
Outputs	Produce two hardcopy reports: 1. Report by pay period 2. Report by month

2. Output Format:

Title	IT Resource Utilization Report - By Project Within Period; By Project Within Month		
Header	None		
Col. Headings	Period, Hours, Cost, Ptype, Project		
Row Content	1 row per period/project; month/project		
Totals/Sub-Totals	Grand Total by Period/Month; Grand Total by Period/Ptype and Month/Ptype		
Footer	None		
Graphics	None		
Other	Format hours with comma and cost as whole U.S. dollar currency		
Orientation:	☒ Portrait ☐ Landscape	Example Attached: ☐ Yes	☒ No

3. Output Logistics:

Medium	☒ Hardcopy	☐ Data	☐ HTML	☐ Other
Frequency	☐ Daily	☒ Monthly	☐ Quarterly	☐ Other
Type	☒ Scheduled	☐ Ad Hoc	☐ User Run	☐ Other
Delivery Location: Hand deliver to R. Powers				
Security/Access Issues: None				

4. Agreement to Proceed:

Estimated Effort: 30 hrs.	Due Date: 5/9/2003
Requester Signature: R. Powers	Date: May 2, 2003

5. Change History:

Date	Requester	Change	Est. Effort	Due Date
5/7/2003	D. Blakeley	Add 2003 rate calculations to increase functionality.	10 hrs.	5/12/2003

Acceptance: _____   Date: _____

Creating SAS Libraries

SAS libraries for this project will follow this structure and will be assigned as follows:

Program Code	Production	'c:\Projects\IT Resource Utilization\code\prod'
	Test	'c:\Projects\IT Resource Utilization\code\test'
Data Sets	Production	'c:\Projects\IT Resource Utilization\data\prod'
	Test	'c:\Projects\IT Resource Utilization\data\test'
User Defined Formats		'c:\Projects\IT Resource Utilization\formats'

Gathering the Data

The data for this project are housed internally, within the same organization. The Requester, in this case the programmer, simply needs to contact another technical resource within the organization to request the Lotus Notes extract. The other two inputs are organizational Master Data to which the programmer already has access. The request is documented by completing the "Input Requirements" template (Figure AIII.3)

When the extract is received and loaded, the Requester updates the "Receipt Information" section of the "Input Requirements" document.

Creating the Analysis Dataset(s)

Reading the Data

RUDBT01A.sas

The first program is a simple mechanism for reading the first few records of a file. In this case we have specified with obs=20 that twenty records should be read.

```
*-----------------------------------------------------------------------*
|                                                                       |
|Program:      RUDBT01A.sas                                             |
|                                                                       |
|Source:       c:\Projects\IT Resource Utilization\code\test            |
|                                                                       |
|Purpose:      Reads and prints the first few records from the Lotus Notes |
|              extract.                                                  |
|                                                                       |
|Input:        c:\work\LNdata\timedata                                  |
|                                                                       |
|Output:       None                                                     |
|                                                                       |
|Update Log:   Notes                              Date       Programmer |
|              -------------------------------    -----------  ------------------ |
|              Original version                   05/12/2003   D. Blakeley |
|                                                                       |
|Usage Notes: The input file is a Lotus Notes view of Period/Name/Project |
|              that was exported from Lotus Notes as tabular text.      |
|                                                                       |
|Other Notes:                                                           |
|                                                                       |
*-----------------------------------------------------------------------*;
data _null_;
   infile 'c:\work\LNdata\timedata' obs=20;
   input;
   put _infile_;
run;
```

147

Figure AIII.3

Input Requirements

Date: 5/1/2003	Project: IT Resource Utilization		
Nature of Request	Extract from the Lotus Notes "Time Tracking System". Please provide a tabular text export of the Period/Name/Project view.		
Date Required: 5/2/2003	Rush:	☒ Yes	☐ No

1. Requester Contact:

Name	D. Blakeley	
Organization	IT Finance	
Postal Address	n/a	
E-mail Address	d.blakeley@anyorg.com	
Telephone: ext. 7826		Fax: n/a

2. Source Business Contact:

Name	Same as #3 below.	
Organization		
Postal Address		
E-mail Address		
Telephone:		Fax:

3. Source Technical Contact:

Name	J. Foster	
Organization	Corporate IT	
Postal Address	n/a	
E-mail Address	j.foster@anyorg.com	
Telephone: ext. 2534		Fax: n/a

4. Data Content/Format:

File Type: ☒ Fixed Field	☐ Free-Format	☐ Hierarchical	☒ Other - variable length records
Delimiter info.			
Time Period	June 1, 2002 through Dec. 31, 2002		
Data Volume	Size: unknown		Records: unknown
Documentation: ☐ Record Layout	☐ Data Model	☐ Data Dictionary ☐ Other	
Medium: ☐ Tape	☐ CD	☐ Hardcopy	☒ Other - diskette

5. Send To:

Name	D. Blakeley
Organization	IT Finance
Delivery Address	n/a

6. Ship Via:

U.S. Mail	☐ First Class	☐ Parcel Post	
UPS	☐ Next Day	☐ 2nd Day	☐ Ground
FedEx	☐ Next Day	☐ 2nd Day	☐ Saturday Delivery
Other	Call me and I will pick it up from your office.		

7. Receipt Information:

Date Received		Received By	
Date Loaded		Loaded By	
Filename(s)		File Location(s)	
Date Reviewed		Reviewed By	

This tool is helpful when dealing with unfamiliar data. The "Input Requirements" document indicates that the Lotus Notes extract is fixed field with variable-length records. Running this program produces the following listing in the SAS log.

```
NOTE: The infile 'c:\work\LNdata\timedata' is:
      File Name=c:\work\LNdata\timedata,
      RECFM=V,LRECL=256

06/15/02
  Allan P. Busher
            Documents - QA Packaging Specs                         0

            International - Asia/Pacific Region Support           10

            International - Europe/Africa Region Support           2

            Loral Support                                          8

            Quality Assurance Support                              4

            Tech Ops - Site Support                               55

(Not Categorized)

  Andrew Friedley
            Field Systems Support and Operations                   8

            Next Genesis Phase II                                 65
NOTE: 20 records were read from the infile 'c:\work\LNdata\timedata'.
      The minimum record length was 0.
      The maximum record length was 79.
NOTE: DATA statement used:
      real time           0.38 seconds
      cpu time            0.05 seconds
```

A review of the SAS log confirms that the record format is variable (RECFM=V) and indicates that the record length is 256 (LRECL=256). We can also see that the fields are fixed, e.g., the name field always starts in column 3. It appears that there are separate records for date, name, and project/hours. In addition, many records are simply blank lines, while others contain "(Not Categorized)". Finally, at the end of the SAS log we see that the maximum record length read by SAS was 79. Based on this information and by manually counting column positions in the SAS log we conclude the following about each field of interest:

Field	Starting Position	Max. Length	Data Type
Period	1	8	Date
Name	3	20	Character
Project	12	60	Character
Hours	78	2	Numeric

We also want to associate each record containing Project and Hours with the appropriate Period and Name. Blank lines and those containing "(Not Categorized)" are of no interest and can be deleted. Based on this initial analysis, we now feel ready to develop a test program that will read the Lotus File extract into a SAS data set.

RUDBR01A.sas

This program is the test version of the code that reads the Lotus Notes tabular text exported view of Period/Name/Project.

```
*--------------------------------------------------------------------*
|                                                                    |
| Program:    RUDBR01A.sas                                           |
|                                                                    |
| Source:     c:\Projects\IT Resource Utilization\code\test          |
|                                                                    |
| Purpose:    Reads the Lotus Notes extract and outputs a temporary SAS data |
|             set and a sample listing of the first 20 records.      |
|                                                                    |
| Input:      c:\work\LNdata\timedata                                |
|                                                                    |
| Output:     RUDBR01A                                               |
|             sample print to output window                          |
|                                                                    |
```

```
| Update Log:  Notes                              Date         Programmer      |
|              ------------------------------     -----------  ---------------- |
|              Original version                   05/12/2003   D. Blakeley      |
|                                                                               |
| Usage Notes: The first attempt to read records from the Lotus Notes extract   |
|              and output to a SAS data set.                                     |
|              Check the SAS Log to ensure that the sum of                       |
|              countblank + countname + countdate + countzero + output          |
|              data set records = total records read.                            |
|                                                                               |
| Other Notes: The input file is a Lotus Notes view of Period/Name/Project      |
|              that was exported from Lotus Notes as tabular text.              |
|                                                                               |
*--------------------------------------------------------------------------*;

* if errors occur print only the first two ;
option errors=2;

data RUDBR01A;

  infile 'c:\work\LNdata\timedata' eof=last obs=2500;

* set the maximum length for the name and project variables to avoid truncation ;
  length name $ 20 project $ 60;

  * retain the values for period and name across records;
  retain period name;

  * the testblank variable reads columns 1-83 and will be used to identify records
    containing all blank, (Not Characterized), date and name values ;
  * the test78 variable reads column 78 will be used to identify records containing
    project hours ;
  * note the use of the trailing @ sign ;
   input @ 1   testblank $char83.
         @78   test78 3. @;

  * identify blank and Not Categorized records ;
  if testblank = ' ' or substr(testblank,1,1) = '(' then do;
    countblank + 1;
    delete;
  end;

  * identify date records ;
  else
  if substr(testblank,3,1) = '/' then do;
    input @1 period mmddyy8.;
    name = ' ';
    project = ' ';
    hours = .;
    countdate + 1;
    delete;
  end;

* identify name records ;
  else
  if substr(testblank,3,1) ne '/' and substr(testblank,3,1) ne ' ' then do;
    input @3 name $20.;
    project = ' ';
    hours = .;
    countname + 1;
    delete;
  end;

* identify records with integer values for project hours in col 78 ;
  else
  if test78 > 0 then do;
    input @12 project $60.
          @78 hr1 6.;
  end;

  * delete records with zero hours reported ;
  if hr1 = . then hr1 = 0;
  if hr2 = . then hr2 = 0;
  hours = hr1+hr2;
  if hours = 0 then do;
    countzero + 1;
    delete;
  end;

  return;
```

```
* print the final value of counter variables in the SAS Log ;
last:
  put countblank= countname= countdate= countzero=;
  stop;
run;

title 'Sample Obs. from RUDBR01A';
proc print data=RUDBR01A (obs=20);
run;
```

Note the amount of coding necessary to account for the different types of records found in a variable-length file format. Each of these unique record formats must be accounted for. In this example, conditional logic is used to classify the input records into those beginning with or containing:

> blank or "(Not Categorized)", else
> > pay period date in the format mm/dd/yy, else
> > > IT resource name, else
> > > > Project description, else
> > > > > Hours, else
> > > > > > zero hours

We also account for how each input record is classified, e.g., "blank", "date", etc. Records that are not relevant to the analysis will not be included in the output data set.

The first attempt to read the Lotus Notes file produces the following note in the SAS log.

```
NOTE: The infile 'c:\work\LNdata\timedata' is:
      File Name=c:\work\LNdata\timedata,
      RECFM=V,LRECL=256

NOTE: Invalid data for test78 in line 15 1-3.
RULE:    ----+----1----+----2----+----3----+----4----+----5----+----6----+----7----+---8----+-
15          (Not Categorized) 17
name=Documents project=  period=. testblank=  test78=. countblank=2 hours=. countdate=0
countname=1 hr1=. hr2=. countzero=2 _ERROR_=1 _N_=5
NOTE: Invalid data for test78 in line 24 1-3.
ERROR: Limit set by ERRORS= option reached. Further errors of this type will not be printed.
24          (Not Categorized) 17
name=Field Syste project=  period=. testblank=  test78=. countblank=3 hours=. countdate=0
countname=2 hr1=. hr2=. countzero=3 _ERROR_=1 _N_=8
countblank=460 countname=93 countdate=1 countzero=278
NOTE: 2500 records were read from the infile 'c:\work\LNdata\timedata'.
      The minimum record length was 0.
      The maximum record length was 83.
NOTE: SAS went to a new line when INPUT statement reached past the end of a line.
```

A review of the SAS log indicates "Invalid data" conditions as well as the following note: "The maximum record layout length was 83." This is a significant finding because it is contrary with our initial assumption that values for the Hours variable were located in columns 78-79. We've discovered that values for Hours can contain up to two decimals and can span up to six columns (i.e., 78-83).

Solution => The SAS System provides the PAD option for use with the Infile statement. It can be very useful when reading variable-length records that contain fixed-field data. The PAD option pads each record with blanks so that all data lines have the same length. The PAD option is useful only when missing data occurs at the end of a record and when SAS encounters an end-of-record marker before the last field is completely read. Adding the PAD option to our Infile statement as shown below and rerunning RUDBR01A.sas solves this problem.

```
infile 'c:\work\LNdata\timedata' eof=last obs=2500 pad;
```

RUDBR01B.sas

This is the production version of the code and is stored in the "prod" folder. We are now ready to read the entire extract file and output a permanent SAS data set. This program contains the following modifications to the test version.

- A permanent SAS data set is output -- note the libname statement.
- The KEEP= option is used to output only the variables relevant to the analysis.
- All input records are read -- note elimination of the "obs=2500" from the Infile statement.
- Only selected variables are printed in the sample listing produces at the end of the program.

```
* output a permanent SAS data set to the production folder ;
libname sasout 'c:\Projects\IT Resource Utilization\data\prod';
run;

* only the four analysis variables will be kept ;
data sasout.RUDBR01B (keep=name period hours project);

   infile 'c:\work\LNdata\timedata' eof=last pad;

.
.
.

* specify the print order of the variables in the report ;
title 'Sample Obs. from RUDBR01B';
proc print data=sasout.RUDBR01B (obs=20);
   var name hours project;
run;
```

Another item of interest in this program is the mechanism for classifying how each record was processed. When reading highly variable data, where a portion of the records are deleted, it is important to account for how each record was handled. The inherent risk with this form of processing is that we will inadvertently fail to read or incorrectly process one or more records.

Recall that in this program we explicitly classified and counted each input record as:

Classification	Counter Variable
Blank or "Not Categorized" record	countblank
Date record	countdate
Name record	countname
Record containing non-zero project hours	n/a
Record containing zero project hours	countzero

The SAS log contains the following note:

```
NOTE: The infile 'c:\work\LNdata\timedata' is:
      File Name=c:\work\LNdata\timedata,
      RECFM=V,LRECL=256

countblank=3187 countdate=11 countname=636 countzero=647
NOTE: 5781 records were read from the infile 'c:\work\LNdata\timedata'.
      The minimum record length was 0.
      The maximum record length was 83.
NOTE: The data set SASOUT.RUDBR01B has 1300 observations and 4 variables.
```

We see that the sum of each record type, plus the number of observations in the output data set equal the number of records that were read.

countblank	3187
countdate	11
countname	636
countzero	647
no. of output observations	1300
no. of records read	5781

We can also see that over half of the input records were blank, which seems consistent with our initial review of the raw data. The number of semi-monthly dates in our analysis period of 6/15/2002-11/15/2002 also corresponds with the value for countdate. Therefore, it appears that we have correctly read in the input data and created a permanent SAS data set. The next step in the process will further validate that assumption.

Validating the Data

RUDBV01A.sas

This program summarizes the data that were output by the previous program. We will compare our results with control totals provided by the data owner. We have been told that there are 43,441.30 hours contained in the extract file. Even though this is only an aggregate total, it gives at least one measure against which we can compare our initial results.

```
*-----------------------------------------------------------------------*
|                                                                       |
|Program:      RUDBV01A.sas                                             |
|                                                                       |
|Source:       c:\Projects\IT Resource Utilization\code\prod            |
|                                                                       |
|Purpose:      Produce summary statistics to compare with control totals.|
|                                                                       |
|Input:        RUDBR01B.sas7bdat                                        |
|                                                                       |
|Output:       Report to be saved from the output window                |
|                                                                       |
|Update Log:   Notes                        Date         Programmer     |
|              -----------------------       ----------   ----------------|
|              Original version             05/12/2003    D. Blakeley    |
|                                                                       |
|Usage Notes:  Compare the value for Hours with an external control report|
|              to validate that the raw data has been processed correctly.|
|                                                                       |
|Other Notes:                                                           |
|                                                                       |
*-----------------------------------------------------------------------*;
libname sasin 'c:\Projects\IT Resource Utilization\data\prod';
run;

title 'Control Total Comparison';
proc tabulate data=sasin.RUDBR01B;
   format period mmddyy8.;
   class period project;
   var hours;
   tables period, hours;
run;
```

This program produces the report displayed below.

Control Total Comparison		
	hours	
	Sum	
period		
06/15/02	3838.25	
06/30/02	3900.75	
07/15/02	3700.75	

153

07/31/02	5457.25
08/15/02	5184.75
08/31/02	4595.00
09/15/02	3900.00
09/30/02	5132.75
10/15/02	5068.00
10/31/02	1524.50
11/15/02	8.00

Totaling this report accounts for only 42,310.00 hours. We are missing 1,131.30 hours. We now have another problem to solve. Where are the missing hours? This is a fairly ominous finding. Fortunately we've discovered this problem early in the process. Including a validation step near the beginning of the process contributes to the identification of errors and omissions sooner rather than later. If problems are found, the recovery time and cost of rework are usually less at this stage then if we were farther into the project.

Additional analysis of the raw data reveals that values for Hours, which we assumed began in column 78, can also begin in column 58. The condition under which occurs seems to be related to the length of the Project Name. In some cases where the Project Name is relatively short, the value for Hours will begin in column 58. However, there are other cases of "short" Project Names where the value for Hours begins in column 78. Therefore, we need to add logic to the program that will identify integer values for Hours beginning in column 78 or column 58. The program version is incremented (new program name is RUDBR01C.sas) after modifying the code as shown below.

```
* the testblank variable reads columns 1-83 and will be used to identify records
    containing all blank, (Not Characterized), date and name values ;
* the test58 and test78 variables read columns 58 and 78 will be used to identify records
    containing project hours ;
* most values for Hours begin in column 78 but in some cases the value for Hours begins
    in col 58 ;
* the error control ?? is used before the test58 informat to generate a missing value
    if the field contains something other than valid numeric data, which would be the
    case if a project description extended to col 58 ;
* note the use of the trailing @ sign ;
input @1  testblank $char83.
      @58 test58 ?? 3.
      @78 test78 3. @;
  .
  .
  .
  .

* typical format -- identify records with integer values for project hours
    in the usual location, which begins in col 78 ;
else
if test78 > 0 then do;
   input @12 project $60.
         @78 hr1 6.;
end;

* nontypical format -- identify records with integer values for project hours
    beginning column 58 ;
else
if test58 > 0 and substr(testblank,57,1)= ' ' then do;
   input @12 project $45.
         @58 hr2 6.;
end;
```

With these changes in place, RUDBR01C.sas is run followed by a rerun of the validation program (RUDBV01A.sas) to verify the new logic and match results to the control totals. A sample listing of the output dataset is produced, a portion of which is shown below

```
                        Sample Obs. from RUDBR01C

   Obs    name                hours    project

     1    Allan P. Busher      10.0    International - Asia/Pacific Region Support
     2    Allan P. Busher       2.0    International - Europe/Africa Region Support
     3    Allan P. Busher       8.0    Loral Support
     4    Allan P. Busher       4.0    Quality Assurance Support
     5    Allan P. Busher      55.0    Tech Ops - Site Support
     6    Andrew Friedley       8.0    Field Systems Support and Operations
     7    Andrew Friedley      65.0    Next Genesis Phase II
```

Combining the Data

RUDBC01A.sas

This is the last program in this step. At this point we are confident that we have correctly read the raw data. We are now ready to merge other data as we continue toward the objective of building the analysis data set. In order to calculate cost, we need to add hourly rate data. The source of this information is one of our "Master Data" files that contains the hourly rate for each IT resource. Since we use "Master Data" such as these over the course of many different projects, we have a much deeper familiarity with the data format.

The "Master Data" file contains two 2002 rates for each person; the regular rate (r02) and the discounted rate (rd02) that will equal the regular rate if no discount is available. The file also contains "place holders" for 2003 rates. Although 2003 rates will not be used in this project, the program is coded to read them so that they can be used in the future if 2003 utilization data were to become available. Finally, the input file also contains a resource type (rtype) code that identifies employees as rtype="E" and contractors as rtype="C".

```
*---------------------------------------------------------------------------*
|                                                                           |
| Program:      RUDBC01A.sas                                                |
|                                                                           |
| Source:       c:\Projects\IT Resource Utilization\code\prod               |
|                                                                           |
| Purpose:      Reads rate data and merges with Period/Name/Project data.   |
|                                                                           |
| Input:        c:\work\Masterdata\rates.dat                                |
|               RUDBR01C.sas7bdat                                           |
|                                                                           |
| Output:       RUDBC01A_RATES.sas7bdat                                     |
|               RUDBC01A.sas7bdat                                           |
|               listing of Master rates data set                            |
|               sample listing to output data set                           |
|                                                                           |
| Update Log:   Notes                      Date          Programmer         |
|               -------------------------  -----------   ------------------ |
|               Original version           05/13/2003    D. Blakeley        |
|                                                                           |
| Usage Notes:  The 2002 rates are valid and are replicated for 2003 as a means |
|               of extending the functionality of this program set.  Update |
|               with actual 2003 rates when they become available.          |
|                                                                           |
| Other Notes:  The input file contains the 2002 rates for each person, where |
|               r02 = gross rates and rd02 = discounted rate.               |
|               Employees are identified as rtype=E and contractors as rtype=C. |
|               The 2002 rates are valid and are replicated for 2003 as a means |
|               of extending the functionality of this program set.         |
*---------------------------------------------------------------------------*;

libname sasout 'c:\Projects\IT Resource Utilization\data\prod';
run;
```

```
* read the Master rate data ;
data sasout.RUDBC01A_RATES;
  infile 'c:\work\Masterdata\rates.dat' lrecl=44 pad;
  input @1  rtype $1.
        @3  name $20.
        @24 r02   5.
        @29 rd02  5.
        @35 r03   5.
        @40 rd03  5.
        ;
run;

title 'Master 2002 Rate Data';
proc print data=sasout.RUDBC01A_RATES;
 format r02 rd02 r03 rd03 dollar7.2;
run;

proc sort data=sasout.RUDBC01A_RATES out=sortrate;
  by name;
run;

proc sort data=sasout.RUDBR01C out=sortr01c;
  by name;
run;

* merge Master rate data with the utilization data ;
data sasout.RUDBC01A;
  merge sortr01c (in=r01c) sortrate;
  by name;
  if r01c;
 run;

title 'Sample 2002 IT Resource Rate Data from RUDBC01A';
proc print data=sasout.RUDBC01A (obs=100);
  var name rtype rd02;
  format rd02 dollar7.2;
run;
```

This program produces two listings: the Master Rates data set and a sample of the merged output data set, portions of which are displayed below.

Master 2002 Rate Data

Obs	rtype	name	r02	rd02	r03	rd03
1	C	Amos R. Paten	$78.00	$78.00	$78.00	$78.00
2	E	Allan P. Busher	.	$88.00	.	$88.00
3	E	Andrew Friedley	.	$64.50	.	$64.50
4	E	Andy K. Manor	.	$88.30	.	$88.30
5	C	Arthur M. Johnsen	$65.00	$65.00	$65.00	$65.00
6	C	Audie Rasmuson	$65.00	$65.00	$65.00	$65.00
7	C	Brian K. Reynolds	$78.00	$78.00	$78.00	$78.00
8	C	Catherine M. Black	$80.00	$80.00	$80.00	$80.00
9	C	D.J. Ross	$48.00	$48.00	$48.00	$48.00
10	C	Dan B. Lemon	$100.00	$92.50	$100.00	$92.50

Sample 2002 IT Resource Rate Data from RUDBC01A

Obs	name	rtype	rd02
50	Allan P. Busher	E	$88.00
51	Allan P. Busher	E	$88.00
52	Allan P. Busher	E	$88.00
53	Allan P. Busher	E	$88.00
54	Allan P. Busher	E	$88.00
55	Allan P. Busher	E	$88.00
56	Allan P. Busher	E	$88.00
57	Allan P. Busher	E	$88.00
58	Amos R. Paten	C	$78.00
59	Amos R. Paten	C	$78.00
60	Andrew Friedley	E	$64.50

Describing the Data

Descriptive Statistics, Outlier Analysis, Data Problems

RUDBD01A.sas

This program produces a variety of descriptive statistics of the merged data set (output from RUDBC01A.sas). A thorough review of these reports can provide more insight into the quality of our data. In general, we want to be able to assess the data in terms of the following three questions:

Are the data:
- complete?
- accurate?
- valid?

The SAS System provides a vast set of tools that can be brought to bear on these questions. We can scrutinize the data to identify missing values, outliers, invalid names, names without rates, rates without names, etc. In addition, we can describe the data in terms of mean, range, and standard deviation for selected variables.

```
*------------------------------------------------------------------*
|                                                                  |
|Program:     RUDBD01A.sas                                         |
|                                                                  |
|Source:      c:\Projects\IT Resource Utilization\code\prod        |
|                                                                  |
|Purpose:     Produces descriptive statistics of the merged data set. |
|                                                                  |
|Input:       RUDBC01A_RATES.sas7bdat                              |
|             RUDBC01A.sas7bdat                                    |
|                                                                  |
|Output:      descriptive statistics to output window              |
|                                                                  |
|Update Log:  Notes                        Date        Programmer  |
|             ---------------------------  ----------  -------------|
|             Original version             05/13/2003  D. Blakeley  |
|                                                                  |
|Usage Notes:                                                      |
|                                                                  |
|Other Notes:                                                      |
|                                                                  |
*------------------------------------------------------------------*;
libname sasin 'c:\Projects\IT Resource Utilization\data\prod';
run;

* distribution of 1996 discounted rates ;
title 'Rates Data Set - RUDBD01A';
proc freq data=sasin.RUDBC01A_RATES;
  tables rd96;
  format rd96 dollar7.2;
run;

* look for missing values, outliers, mean, range, etc. ;
title 'Rates Data Set - RUDBD01A';
proc univariate data=sasin.RUDBC01A_RATES;
  var rd96;
run;

* look for invalid names ;
title 'From RUDBD01A';
proc freq data=sasin.RUDBC01A;
  tables name / nopercent norow nocol;
run;

title 'Names without Rates or Rates without Names';
data _null_;
  set sasin.RUDBC01A end=last;
  if name = ' ' or rd96 = . then do;
    put _n_= name= rd96=;
```

```
      counter + 1;
   end;
   else if last and counter=0 then do;
      put 'No missing names or rates in the data set';
   end;
run;
```

Note the section of code near the bottom of the program that evaluates each record to identify any occurrences of missing name or rate values. If this condition is encountered, the program will print the observation number, name and rate to the SAS log. Otherwise the program will print a 'No missing names or rates in the data set' message to the SAS log.

Portions of the report output are displayed below.

```
                    Rates Data Set - RUDBD01A

                       The FREQ Procedure

                                       Cumulative    Cumulative
     rd02     Frequency     Percent     Frequency      Percent
   -----------------------------------------------------------
   $22.00         1          1.32           1           1.32
   $35.00         1          1.32           2           2.63
   $36.00         2          2.63           4           5.26
   $44.00         1          1.32           5           6.58
   $48.00         2          2.63           7           9.21
   $50.00         2          2.63           9          11.84
   $52.00         1          1.32          10          13.16
   $55.00         1          1.32          11          14.47
   $60.00         4          5.26          15          19.74
   $60.10         1          1.32          16          21.05
      .
      .
      .
  $100.00         2          2.63          69          90.79
  $110.00         1          1.32          70          92.11
  $112.50         1          1.32          71          93.42
  $117.00         3          3.95          74          97.37
  $118.00         2          2.63          76         100.00
```

The lowest and highest values in the frequency distribution of 2002 discounted rates are displayed above. The values in this report should be reviewed for reasonableness. Do you expect to see hourly rates below \$40.00? This should trigger some investigation into the validity of those rates. As it turned out, these "low" rates were associated with an intern, one non-exempt project assistant, and an entry-level programmer.

```
                    The UNIVARIATE Procedure
                       Variable:  rd02


                    Basic Statistical Measures

          Location                      Variability

   Mean      73.62237    Std Deviation          20.17520
   Median    71.55000    Variance              407.03883
   Mode      65.00000    Range                  96.00000
                         Interquartile Range    21.70000


                 Tests for Location: Mu0=0

       Test           -Statistic-      -----p Value------

   Student's t      t   31.81256    Pr > |t|    <.0001
   Sign             M         38    Pr >= |M|   <.0001
   Signed Rank      S       1463    Pr >= |S|   <.0001
```

```
                    Quantiles (Definition 5)

                    Quantile        Estimate

                    100% Max         118.00
                    99%              118.00
                    95%              117.00
                    90%              100.00
                    75% Q3            84.70
                    50% Median        71.55
                    25% Q1            63.00
                    10%               50.00
                    5%                36.00
                    1%                22.00
                    0% Min            22.00

                      Extreme Observations

          ----Lowest----             ----Highest---

          Value         Obs          Value        Obs

             22          32            117          21
             35          43            117          74
             36          66            117          75
             36          18            118          12
             44          53            118          49
```

A portion of the output from the Univariate procedure is displayed above. A wealth of information is available in this report. A review of the mean ($73.62), standard deviation ($20.17) and range ($96.00) appear to be within expectations. The "Extreme Observations" section of the report lists the five lowest and highest values for the 2002 discounted rate. This is simply another view of the same information obtained from the frequency distribution of rates, and is a useful indicator of the presence of outliers in the data set.

```
                        The FREQ Procedure

                                                  Cumulative
            name                   Frequency      Frequency

            Allan P. Busher            57              57
            Amos R. Paten               2              59
            Andrew Friedley            18              77
            Andy K. Manor              45             122
            Arthur M. Johnsen          10             132
            Audie Rasmuson             18             150
            Brian K. Reynolds           9             159
            Catherine M. Black         14             173
            D.J. Ross                  10             183
            Dan B. Lemon                6             189
```

A portion of the frequency distribution of IT resource name is displayed above. This report should be reviewed to identify any misspelling of Name, which would have adversely affected the merge by Name in RUDBC01A.sas.

Finally, we can review the SAS log to look for any evidence of unmatched Name/Rate combinations.

```
490   title 'Names without Rates or Rates without Names';
491   data _null_;
492     set sasin.RUDBC01A end=last;
493     if name = ' ' or rd96 = . then do;
494       put _n_= name= rd96=;
495       counter + 1;
496     end;
497     else if last and counter=0 then do;
498       put 'No missing names or rates in the data set';
499     end;
500   run;

No missing names or rates in the data set
```

```
NOTE: There were 1379 observations read from the data set SASIN.RUDBC01A.
NOTE: DATA statement used:
      real time              0.03 seconds
      cpu time               0.03 seconds
```

The SAS log note 'No missing names or rates in the data set' indicates that each name has a rate and vice versa.

Based on our review we have concluded that the data set is complete, accurate, and valid. We are now ready to manipulate the data into the final form to support our analysis.

Answering the Questions

Manipulating the Data

<u>RUDBM01A.sas</u>

One of our report requirements is to classify each project into a "Business Change" or "Support" type. The most recently created data set, RUDBC01A, does not contain a project type variable. However, we do have access to another "Master Data" file that associates a value for project type with each project name. We need to devise a method for linking, merging, or otherwise looking up the project type that corresponds to each of the projects in RUDBC01A. There are several methods for accomplishing this task. I have chosen to use a "table look-up" approach that will create a user defined format that associates each project with its project type.

```
*---------------------------------------------------------------------*

 Program:     RUDBM01A.sas

 Source:      c:\Projects\IT Resource Utilization\code\prod

 Purpose:     Create two formats for use in RUDBM02A.sas

 Input:       c:\work\Masterdata\project.dat

 Output:      character format $ptype, which represents the project type
              'B' for Business Change and 'S' for Support
              format mo, which represents the months of June-Dec 1996

 Update Log:  Notes                               Date         Programmer
              -----------------------------       -----------  ------------------
              Original version                    05/13/2003   D. Blakeley

 Usage Notes:

 Other Notes: The formats are stored in the format library specified by the
              libname statement below.

*---------------------------------------------------------------------*;

libname library 'c:\Projects\IT Resource Utilization\formats';
run;

* create a permanent user defined format for project type (ptype) ;
data ptype (keep=ptype fmtname start label type hlo);
   infile 'c:\work\Masterdata\project.dat' lrecl= 73 pad;
   retain fmtname 'ptype';
   length start $ 60;
   input @1  ptype      $1.
         @3  project    $60.;
   start = project;
   label = ptype;
   type  = 'c';
   hlo   = ' ';
   output;
run;

proc format library=library
   cntlin=ptype (keep=fmtname start label type hlo);
```

```
run;

* create a permanent user defined format for month (mo) ;
proc format library=library;
  value mo  0602 = 'Jun 02'
            0702 = 'Jul 02'
            0802 = 'Aug 02'
            0902 = 'Sep 02'
            1002 = 'Oct 02'
            1102 = 'Nov 02'
            1202 = 'Dec 02'
                 ;
run;
```

Note that the format ($ptype) is permanently stored and can be called from the program that will create the project type variable. In this program we also create a second permanent user defined format (mo) which be used to format date values (mmyy) for reporting purposes.

RUDBM02A.sas

This program pulls everything together and calculates the individual activity cost for each record in the data set. This calculation varies depending on whether the IT resource is an employee or contractor. It also varies depending on the time period in which the activity occurred (i.e., CY 2002 or CY 2003). In this program we also calculate a new variable 'month', which indicates the month/year in which the activity occurred. The 'month' variable will allow us to satisfy the requirement that we report hours and cost on a monthly basis, in addition to a pay period basis.

```
*-----------------------------------------------------------------------*
|                                                                       |
|Program:     RUDBM02A.sas                                              |
|                                                                       |
|Source:      c:\Projects\IT Resource Utilization\code\prod             |
|                                                                       |
|Purpose:     Calculates activity cost and month that the cost was charged |
|                                                                       |
|Input:       RUDBC01A.sas7bdat                                         |
|             format catalog                                            |
|                                                                       |
|Output:      RUDBM02A.sas7bdat (analysis data set)                     |
|             sample listing to output window                           |
|                                                                       |
|Update Log:  Notes                       Date          Programmer      |
|             ---------------------------  -----------   ---------------- |
|             Original version             05/13/2003    D. Blakeley     |
|                                                                       |
|Usage Notes:                                                           |
|                                                                       |
|Other Notes: Contractor costs are the product of hours and hourly rate. |
|             Employee cost calculations are more complex.  Each employee |
|             has an hourly rate proxy, which is a function of annual salary. |
|             The proxy rate is annualized by multiplying by 1000 hours and |
|             dividing by 2080 hours.  The result is multiplied by the   |
|             overhead loading factor (1.32 for 1996) and then multiplied by |
|             hours.                                                    |
|             Processing logic for 1997 is included even though 1997 rates |
|             are not available.                                        |
*-----------------------------------------------------------------------*;

libname sasin 'c:\Projects\IT Resource Utilization\data\prod';
run;

libname library 'c:\Projects\IT Resource Utilization\formats';
run;

data sasin.RUDBM02A;
  set sasin.RUDBC01A;

  * use the $ptype format to calculate a value for ptype ;
  ptype = put(project,$ptype.);

  * calculate activity cost using the appropriate hourly rate
    and the hours charged by each individual IT resource ;
```

```
   if period <= '31DEC02'd then do;
     if rtype = 'C' then do;
       cost = hours*rd02;
     end;
     else
     if rtype = 'E' then do;
       cost = hours*(((rd02*1000)/2080)*1.32);
     end;
   end;
   else
   if period > '31DEC02'd then do;
     if rtype = 'C' then do;
       cost = hours*rd03;
     end;
     else
     if rtype = 'E' then do;
       cost = hours*(((rd03*1000)/2080)*1.40);
     end;
   end;

* calculate a value for month that will sort properly ;
   select;
     when(period='15JUN02'd, period='30JUN02'd) do;
       month = 062002;
     end;
     when(period='15JUL02'd, period='31JUL02'd) do;
       month = 072002;
     end;
     when(period='15AUG02'd, period='31AUG02'd) do;
       month = 082002;
     end;
     when(period='15SEP02'd, period='30SEP02'd) do;
       month = 092002;
     end;
     when(period='15OCT02'd, period='31OCT02'd) do;
       month = 102002;
     end;
     when(period='15NOV02'd, period='30NOV02'd) do;
       month = 112002;
     end;
     when(period='15DEC02'd, period='31DEC02'd) do;
       month = 122002;
     end;
     otherwise month = 999999;
   end;
run;

title 'Sample Obs. from RUDBM02A';
proc print data=sasin.RUDBM02A (obs=100);
  var name rtype hours rd02 cost;
run;
```

The dataset output from this program is the "Analysis Dataset" from which we will do all the final reporting. A portion of that data set is displayed below.

```
                    Sample Obs. from RUDBM02A

     Obs          name           rtype     hours     rd02       cost

      50     Allan P. Busher        E       22.0     88.0     1228.62
      51     Allan P. Busher        E        7.0     88.0      390.92
      52     Allan P. Busher        E        9.0     88.0      502.62
      53     Allan P. Busher        E       25.0     88.0     1396.15
      54     Allan P. Busher        E        2.0     88.0      111.69
      55     Allan P. Busher        E        1.0     88.0       55.85
      56     Allan P. Busher        E        7.0     88.0      390.92
      57     Allan P. Busher        E       11.0     88.0      614.31
      58     Amos R. Paten          C        8.0     78.0      624.00
      59     Amos R. Paten          C        8.0     78.0      624.00
      60     Andrew Friedley        E        8.0     64.5      327.46
```

In this "Analysis Dataset" we have all the elements necessary to deliver the reports as specified.

Reporting

RUDBP01A.mac

The "Output Requirements" document indicates that our primary objective for this project is to answer the following question: "What are the labor hours and estimated cost incurred for each project calculated on a per pay period basis and monthly basis?". The only thing that varies is the level of analysis, i.e., pay period or monthly. Rather than write two reporting programs that would be nearly identical, we can write a single macro that contains all the common code but allows us to substitute parameters that determine whether the report will be produced on a per pay period basis or a monthly basis. We potentially save a modest amount of coding time, but more importantly, having fewer individual programs can reduce the ongoing maintenance effort.

```
%macro report(duration,format);
%*----------------------------------------------------------------------*
|
| Program:     RUDBP01A.mac
|
| Source:      c:\Projects\IT Resource Utilization\code\prod
|
| Purpose:     Produce report of project hours and costs by selected duration
|
| Input:       RUDBM02A.sas7bdat
|
| Output:      listings to output window
|
| Update Log:  Notes                              Date         Programmer
|              --------------------------------   ----------   ------------
|              Original version                   05/14/2003   D. Blakely
|
| Usage Notes:
|
| Other Notes:
|
*----------------------------------------------------------------------*;
title1 'IT Resource Utilization Report';
title2 'Produced by RUDBP01A';

%* summarize across all records ;
proc sql:
   title3 "Grand Total By &duration";
   select &duration "&duration" format &format,
          sum(hours) as Hours format comma9.2,
          sum(cost) as Cost format dollar8.
   from sasin.RUDBM02A
   group by &duration;
quit;

%* summarize by Ptype ;
proc sql:
   title3 'Grand Total By Ptype';
   select &duration "&duration" format &format,
          sum(hours) as Hours format comma9.2,
          sum(cost) as Cost format dollar8.,
          ptype 'Ptype'
   from sasin.RUDBM02A
   group by &duration, ptype;
quit;

%* summarize by project within specified duration ;
%* shorten the length of the Project description
   so that it will fit on the same page ;
proc sql noprint;
   create table project_within_duration_report as
   select &duration "&duration" format &format,
          sum(hours) as Hours format comma9.2,
          sum(cost) as Cost format dollar8.,
          ptype 'Ptype',
          substr(project,1,50) as Project
   from sasin.RUDBM02A
   group by &duration, project, ptype;
quit;
```

```
%* use proc report to format the output so that the
    duration appears only once for each project group ;
proc report data=project_within_duration_report
            headline
            headskip
            nowd;
  title3 "By Project Within &duration";
  column &duration hours cost ptype project;
  define &duration / order order=internal;
  define ptype     / width=5;
  define project   / width=50;
run;

%mend report;
```

Note that the order=internal option was used in the PROC REPORT define statement. This option orders values by their unformatted values, which yields the same order that PROC SORT would yield. PROC REPORT uses formatted values as the default sort order, which in the case of this report would have resulted in a non-chronological sequence. See the Issue Log at the end of the case for more information.

RUDBP01A.sas

This program calls the report macro above: once to report by pay period and again to report by month. Formats for period and month are also specified when calling the report macro.

```
*--------------------------------------------------------------------------*
|                                                                          |
|Program:      RUDBP01A.sas                                                |
|                                                                          |
|Source:       c:\Projects\IT Resource Utilization\code\prod               |
|                                                                          |
|Purpose:      Produce report of project hours and costs by the selected   |
|              duration, either Period or Month                            |
|                                                                          |
|Input:        RUDBP01A.mac                                                |
|              RUDBM02A.sas7bdat                                           |
|                                                                          |
|Output:       listings to output window                                   |
|                                                                          |
|Update Log:   Notes                           Date        Programmer      |
|              ------------------------------   ----------- ---------------- |
|              Original version                 05/14/2003  D. Blakely      |
|                                                                          |
|Usage Notes:  This program calls the report macro that has the following  |
|              two macro parameters and their associated values:           |
|                  1. duration (Period or Month)                           |
|                  2. format (mmddyy8. or mo.)                             |
|                                                                          |
|Other Notes:                                                              |
|                                                                          |
|                                                                          |
*--------------------------------------------------------------------------*;

libname sasin 'c:\Projects\IT Resource Utilization\data\prod';
run;

libname library 'c:\Projects\IT Resource Utilization\formats';
run;

%include 'c:\Projects\IT Resource Utilization\code\prod\RUDBP01A.mac';

%report(Period,mmddyy8.)
%report(Month,mo.)
```

A portion of each report is displayed below.

```
                    IT Resource Utilization Report
                         Produced by RUDBP01A
                         Grand Total By Period

                   Period        Hours        Cost
                   ────────────────────────────────
                   06/15/02    3,840.25    $231,539
                   06/30/02    3,963.25    $238,134
                   07/15/02    3,786.25    $231,929
                   07/31/02    5,600.25    $356,405
                   08/15/02    5,322.75    $344,668
                   08/31/02    4,750.30    $294,932
                   09/15/02    4,019.50    $259,520
                   09/30/02    5,349.25    $333,470
                   10/15/02    5,277.00    $330,842
                   10/31/02    1,524.50     $82,980
                   11/15/02        8.00        $624


                    IT Resource Utilization Report
                         Produced by RUDBP01A
                         Grand Total By Ptype

                   Period        Hours        Cost    Ptype
                   ────────────────────────────────────────
                   06/15/02    1,471.00     $93,563     B
                   06/15/02    2,369.25    $137,976     S
                   06/30/02    1,397.50     $81,725     B
                   06/30/02    2,565.75    $156,409     S
                   07/15/02    1,315.75     $79,765     B
                   07/15/02    2,470.50    $152,164     S
                   07/31/02    2,096.00    $143,557     B
                   07/31/02    3,504.25    $212,849     S
                   08/15/02    2,219.25    $153,478     B
                   08/15/02    3,103.50    $191,189     S
                   08/31/02    1,959.80    $125,491     B
                   08/31/02    2,790.50    $169,441     S
                   09/15/02    1,699.75    $114,934     B
                   09/15/02    2,319.75    $144,586     S
                   09/30/02    2,272.00    $143,950     B
                   09/30/02    3,077.25    $189,519     S
                   10/15/02    2,221.00    $149,232     B
                   10/15/02    3,056.00    $181,609     S
                   10/31/02      515.50     $26,874     B
                   10/31/02    1,009.00     $56,106     S
                   11/15/02        8.00        $624     S


                    IT Resource Utilization Report
                         Produced by RUDBP01A
                         By Project Within Period

Period        Hours       Cost   Ptype   Project
────────────────────────────────────────────────────────────────────────
06/15/02      65.00     $4,654     B     ACT Domain Reporting - Phase 1
              35.00     $1,471     B     CABS - Enhancements
              42.50     $3,088     B     CABS/IS Upgrade
               4.00       $226     B     Contract Vendor database
             191.00    $13,240     B     Data Access - Phase I
              45.00     $3,370     B     GBU Implementation Planning (Project Quantify)
              16.00       $666     B     Global Business Research & I


                    IT Resource Utilization Report
                         Produced by RUDBP01A
                         Grand Total By Month

                   Month        Hours        Cost
                   ────────────────────────────────
                   Jun 02     7,803.50    $469,672
                   Jul 02     9,386.50    $588,334
                   Aug 02    10,073.05    $639,600
                   Sep 02     9,368.75    $592,990
                   Oct 02     6,801.50    $413,821
                   Nov 02         8.00        $624
```

```
                    IT Resource Utilization Report
                         Produced by RUDBP01A
                         Grand Total By Ptype

              Month       Hours        Cost   Ptype
              ─────────────────────────────────────
              Jun 02    2,868.50   $175,287   B
              Jun 02    4,935.00   $294,385   S
              Jul 02    3,411.75   $223,322   B
              Jul 02    5,974.75   $365,013   S
              Aug 02    4,179.05   $278,969   B
              Aug 02    5,894.00   $360,630   S
              Sep 02    3,971.75   $258,884   B
              Sep 02    5,397.00   $334,106   S
              Oct 02    2,736.50   $176,106   B
              Oct 02    4,065.00   $237,715   S
              Nov 02        8.00       $624   S


                    IT Resource Utilization Report
                         Produced by RUDBP01A
                        By Project Within Month

Month      Hours       Cost   Ptype Project
──────────────────────────────────────────────────────────────────────────────
Jun 02    109.00     $7,967   B     ACT Domain Reporting - Phase 1
           45.00     $1,884   B     CABS - Enhancements
          168.00     $9,736   B     CABS/IS Upgrade
            8.00       $452   B     Contract Vendor database
          515.50    $32,940   B     Data Access - Phase I
           75.00     $5,616   B     GBU Implementation Planning (Project Quantify)
           32.00     $1,276   B     Global Business Research & Intelligence
           23.00     $1,533   B     IT - Process Improvement
           22.00     $1,647   B     IT Priority Setting Process
          271.00    $15,390   B     Lotum Databases
          277.50    $17,814   B     Managers Analysis Tool
          163.00     $9,685   B     Next Genesis - Interim Releases
           53.00     $5,099   B     Next Genesis - Organization Reference Database , O
          577.50    $37,805   B     Next Genesis Phase II
            8.00       $387   B     Opportunity Pipeline
           43.00     $2,373   B     PRISO Implementation
           90.00     $4,190   B     Product Reference Database Phase I
          244.00    $12,597   B     S/P Financials - Europe
          144.00     $6,897   B     S/P Financials - North America
          531.00    $29,835   S     Central Managed Care (Contract Administration) Sup
          211.00     $9,939   S     Corporate Finance Support
          951.50    $58,389   S     Field Systems Support and Operations
```

Staying Organized

Building the Audit Trail

Using a program run log was useful in this project. Listed below is a partial extract that lists the chronological sequence of initial events that were critical to understanding the raw data format and resolving the issues that arose. The log distinguishes test runs from production runs, includes comments that are helpful in reconstructing events, and otherwise captures information that can come in handy when sorting through output and other documentation. For example, the RUDBR01B program, which initially appeared to be a valid production run that read the entire Lotus Notes raw data file, turned out to inadvertently exclude a number of records. It was replaced by RUDBR01C, which become the final version. Therefore, any output that was associated with RUDBR01B becomes irrelevant and can be discarded from final project documentation.

Date	Program	Type	Comments	Results
5/12/03	RUDBT01A	T	obs=20	max rec length=79
5/12/03	RUDBR01A	T	obs=2500	invalid data; max rec length=83
5/12/03	RUDBR01A	T	PAD option	O.K.
5/12/03	RUDBR01B	P		looks O.K.
5/12/03	RUDBV01A	P		control totals don't match
5/12/03	RUDBR01C	P	add logic for col. 58 and col. 78	
5/12/03	RUDBV01A	P	no output saved	rerun to verify new logic
5/13/03	RUDBC01A	P		O.K.
5/13/03	RUDBD01A	P		O.K.
5/13/03	RUDBM01A	P		O.K.
5/13/03	RUDBM02A	P		O.K.
5/14/03	RUDBP01A	P		O.K.

You can see from the file dates in the project folders below that some of the programs were subsequently rerun.

Other steps taken during the course of the project include:

- Saving interim steps in the form of data sets or report listings. See the flowchart below.
- Updating the documentation block for each program to accurately reflect inputs, outputs, and changes.
- Ensuring that the programs are adequately commented.
- Saving test results that document the accuracy of your programming.

Managing Files

The folder configuration displayed in Figure AIII.2 was used to organize and file all project-specific code, data, documentation, and formats. Note that "Master Data" files, which are not project-specific, are not duplicated in the project folders.

Folder contents at the conclusion of the project are displayed below.

Projects\IT Resource Utilization\code\prod

Name	Size	Type	Modified
RUDBC01A.sas	4 KB	SAS File	5/27/03 3:18 PM
RUDBD01A.sas	3 KB	SAS File	5/27/03 3:25 PM
RUDBM01A.sas	3 KB	SAS File	5/15/03 1:25 PM
RUDBM02A.sas	4 KB	SAS File	5/15/03 1:30 PM
RUDBP01A.mac	4 KB	MAC File	5/20/03 2:31 PM
RUDBP01A.sas	3 KB	SAS File	5/15/03 1:42 PM
RUDBR01C.sas	6 KB	SAS File	5/27/03 3:09 PM
RUDBV01A.sas	2 KB	SAS File	5/15/03 12:27 PM

Projects\IT Resource Utilization\code\test

Name	Size	Type	Modified
RUDBR01A.sas	5 KB	SAS File	5/27/03 2:43 PM
RUDBT01A.sas	2 KB	SAS File	5/14/03 4:11 PM
Program Doc Template.txt	2 KB	Text Document	5/13/03 2:49 PM

Projects\IT Resource Utilization\data\prod

Name	Size	Type	Modified
rudbc01a.sas7bdat	193 KB	SAS7BDAT File	5/27/03 3:19 PM
rudbc01a_rates.sas7bdat	9 KB	SAS7BDAT File	5/27/03 3:19 PM
rudbm02a.sas7bdat	217 KB	SAS7BDAT File	5/19/03 10:13 AM
rudbr01c.sas7bdat	137 KB	SAS7BDAT File	5/27/03 3:10 PM

Projects\IT Resource Utilization\data\test

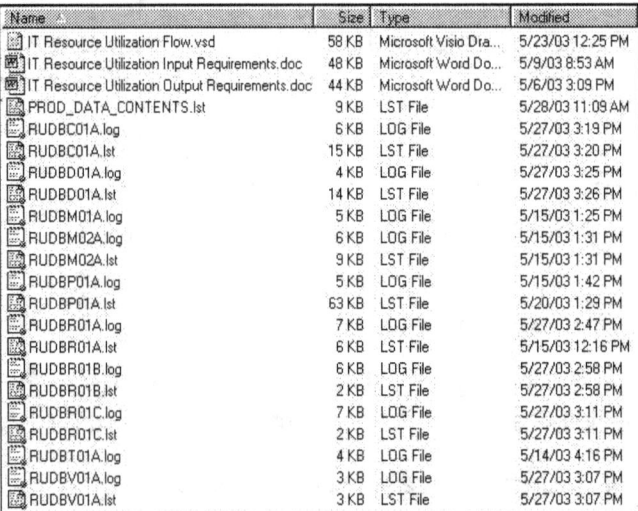

Projects\IT Resource Utilization\doc

Name	Size	Type	Modified
IT Resource Utilization Flow.vsd	58 KB	Microsoft Visio Dra...	5/23/03 12:25 PM
IT Resource Utilization Input Requirements.doc	48 KB	Microsoft Word Do...	5/9/03 8:53 AM
IT Resource Utilization Output Requirements.doc	44 KB	Microsoft Word Do...	5/6/03 3:09 PM
PROD_DATA_CONTENTS.lst	9 KB	LST File	5/28/03 11:09 AM
RUDBC01A.log	6 KB	LOG File	5/27/03 3:19 PM
RUDBC01A.lst	15 KB	LST File	5/27/03 3:20 PM
RUDBD01A.log	4 KB	LOG File	5/27/03 3:25 PM
RUDBD01A.lst	14 KB	LST File	5/27/03 3:26 PM
RUDBM01A.log	5 KB	LOG File	5/15/03 1:25 PM
RUDBM02A.log	6 KB	LOG File	5/15/03 1:31 PM
RUDBM02A.lst	9 KB	LST File	5/15/03 1:31 PM
RUDBP01A.log	5 KB	LOG File	5/15/03 1:42 PM
RUDBP01A.lst	63 KB	LST File	5/20/03 1:29 PM
RUDBR01A.log	7 KB	LOG File	5/27/03 2:47 PM
RUDBR01A.lst	6 KB	LST File	5/15/03 12:16 PM
RUDBR01B.log	6 KB	LOG File	5/27/03 2:58 PM
RUDBR01B.lst	2 KB	LST File	5/27/03 2:58 PM
RUDBR01C.log	7 KB	LOG File	5/27/03 3:11 PM
RUDBR01C.lst	2 KB	LST File	5/27/03 3:11 PM
RUDBT01A.log	4 KB	LOG File	5/14/03 4:16 PM
RUDBV01A.log	3 KB	LOG File	5/27/03 3:07 PM
RUDBV01A.lst	3 KB	LST File	5/27/03 3:07 PM

Projects\IT Resource Utilization\formats

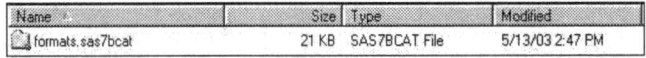

Managing Change

The only change in scope encountered in this project was to add functionality to the cost calculation program (RUDBM02A.sas) by including processing logic to accommodate 2003 rates. Even though the data for this project was limited to the 2002 timeframe, the decision was made to further extend the potential use of this program by adding the capability to use future rates, in the event the transaction data become available. Note that this change, which included a revision to the project due date, was recorded in the "Output Requirements" form displayed in Figure AIII.1.

Documenting the Results

Making the Results Replicable

A review of the folder contents above shows that for each program we have saved the following information:
- Final code (in the code\prod folder)
- SAS logs (in the doc folder)
- Sample prints (in the doc folder)

The CONTENTS and DATASETS Procedures

We have also used the CONTENTS procedure to document each production SAS data set using the code displayed below.

```
libname sasin 'c:\Projects\IT Resource Utilization\data\prod';
proc contents data=sasin._all_;
run;
```

Partial output from the CONTENTS procedure is displayed below

Directory Listing of the data\prod Folder

```
                        The SAS System

                      The CONTENTS Procedure

                        -----Directory-----

      Libref:            SASIN
      Engine:            V8
      Physical Name:     c:\Projects\IT Resource Utilization\data\prod
      File Name:         c:\Projects\IT Resource Utilization\data\prod


                                     File
      #   Name             Memtype   Size      Last Modified
      _____

      1   RUDBC01A         DATA      197632    27MAY2003:15:18:59
      2   RUDBC01A_RATES   DATA      9216      27MAY2003:15:18:59
      3   RUDBM02A         DATA      222208    19MAY2003:10:13:19
      4   RUDBR01C         DATA      140288    27MAY2003:15:10:23
```

Contents of the "Analysis Dataset" RUDBM02A.sas7bdat

```
                        The CONTENTS Procedure

      Data Set Name:  SASIN.RUDBM02A              Observations:          1379
      Member Type:    DATA                        Variables:             12
      Engine:         V8                          Indexes:               0
      Created:        13:30 Thursday, May 15, 2003 Observation Length:   152
      Last Modified:  13:30 Thursday, May 15, 2003 Deleted Observations: 0
      Protection:                                 Compressed:            NO
      Data Set Type:                              Sorted:                NO
      Label:


              -----Engine/Host Dependent Information-----

      Data Set Page Size:           12288
      Number of Data Set Pages:     18
      First Data Page:              1
      Max Obs per Page:             80
      Obs in First Data Page:       66
      Number of Data Set Repairs:   0
      File Name:                    c:\Projects\IT Resource Utilization\data\prod\rudbm02a.sas7bdat
      Release Created:              8.0202M0
      Host Created:                 WIN_PRO


              -----Alphabetic List of Variables and Attributes-----

              #    Variable    Type    Len    Pos
              _____

              11   cost        Num      8     48
              4    hours       Num      8      8
              12   month       Num      8     56
              1    name        Char    20     64
              3    period      Num      8      0
              2    project     Char    60     84
              10   ptype       Char     1    145
              6    r96         Num      8     16
              8    r97         Num      8     32
              7    rd96        Num      8     24
              9    rd97        Num      8     40
              5    rtype       Char     1    144
```

169

The CONTENTS procedure output should also be saved in the doc folder.

Flowcharting

The project flowchart was prepared to document each input, program, and output. It is displayed in Figure AIII.4.

Keeping Track of Problems Solved

This project presented several interesting technical challenges. Each is captured in an "Issue Log" and filed with the project documentation. Equally important is to make issue log information available to other analysts who may benefit from what was learned in the course of this project.

The Issue Logs displayed at the end of the chapter cover the following topics:
- PAD infile option for reading the Lotus Notes extract
- Values for project hours that can begin in column 58 or column 78
- PROC REPORT sort order
- Format creation technique

Final Project Archiving

All of the files contained in the project folders should be copied for storage on a separate device and/or physical location.

Figure AIII.4

IT Resource Utilization

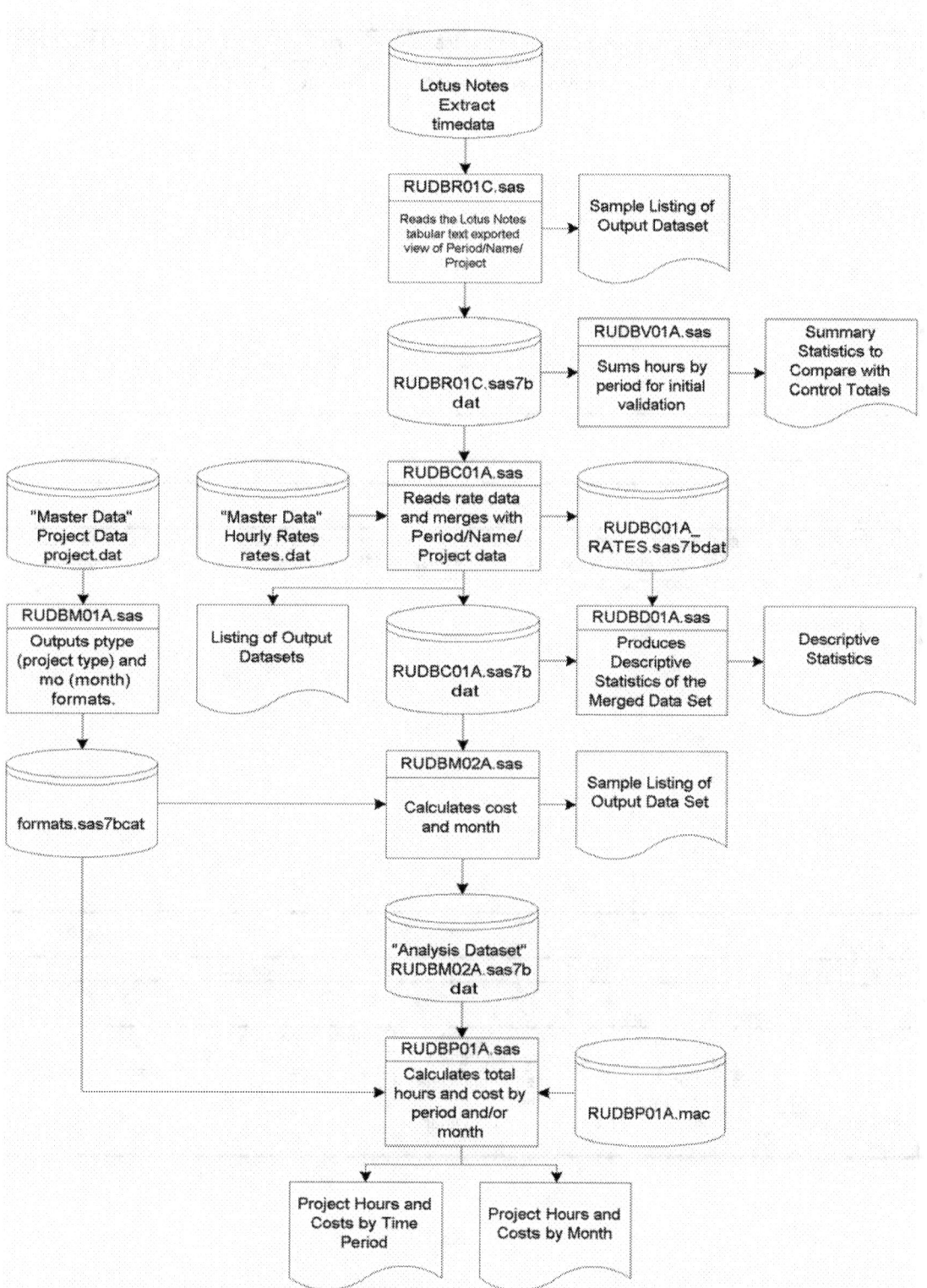

Issue Log - PAD Infile Option For Reading the Lotus Notes Extract

Date: 5/12/03	Project: IT Resource Utilization
Initiator Name	D. Blakeley

1. Description:

The original expectation (actually an assumption) was that the values for the Hours variable were located in columns 78-79. I discovered that values for Hours can contain up to two decimals and can span up to six columns (i.e., 78-83). What we have here is a situation of variable-length records that contain fixed-field data.

2. Turnover:

Issue Passed To: File	Date: 5/13/03

3. Resolution:

Use the PAD option with the infile statement. The PAD option pads each record with blanks so that all data lines have the same length. The PAD option is useful only when missing data occurs at the end of a record and when SAS encounters an end-of-record marker before the last field is completely read.

This issue is resolved.

4. Return:

Issue Received From: n/a	Date: n/a

5. Classification - check all that apply:

Data	☐ Missing Values	☐ Wrong Dataset	☐ Incomplete Data
	☒ Different Format	☐ Unknown Values	☐ Other
Coding	☐ Faulty Logic	☐ Used Wrong Data	☐ Formula Error
	☐ Selection Error	☐ Other	
Infrastructure	☐ Memory	☐ Space	☐ Other
Other			

Issue Log - Values For Project Hours That Can Begin in Column 58 Or Column 78

Date: 5/12/03	Project: IT Resource Utilization
Initiator Name	D. Blakeley

1. Description:

Another assumption early in this project was that all values for Hours began in column 78. When the initial comparison to control report totals didn't match, I discovered that I was missing some records and therefore under reporting total Hours.

Additional analysis of the raw data revealed that value for Hours could also begin in column 58. The condition under which this occurs seems to be related to the length of the Project Name. In some case where the Project Name is relatively short, the value for Hours will begin in column 58. However, there are other cases of "short" Project Names where the value for Hours begins in column 78.

2. Turnover:

Issue Passed To: File	Date: 5/13/03

3. Resolution:

Logic was added to the program that will identify integer values for Hours beginning in column 78 or column 58.

4. Return:

Issue Received From: n/a	Date: n/a

5. Classification - check all that apply:

Data	☐ Missing Values	☐ Wrong Dataset	☐ Incomplete Data
	☒ Different Format	☐ Unknown Values	☐ Other
Coding	☐ Faulty Logic	☐ Used Wrong Data	☐ Formula Error
	☐ Selection Error	☐ Other	
Infrastructure	☐ Memory	☐ Space	☐ Other
Other			

Issue Log - PROC REPORT Sort Order

Date: 5/14/03	Project: IT Resource Utilization
Initiator Name	D. Blakeley

1. Description:

The PROC REPORT output from RUDBP01A.sas was sorting by the formatted value when reporting by Month. The formatted values are 'Jun 02', 'Jul 02', 'Aug 02', etc. This resulted in a non-chronological sequence.

2. Turnover:

Issue Passed To: File	Date: 5/15/03

3. Resolution:

Use the order=internal option in the PROC REPORT define statement. This option orders values by their unformatted values, which yields the same order that PROC SORT would yield.

4. Return:

Issue Received From: n/a	Date: n/a

5. Classification - check all that apply:

Data	☐ Missing Values ☐ Wrong Dataset ☐ Incomplete Data
	☐ Different Format ☐ Unknown Values ☐ Other
Coding	☐ Faulty Logic ☐ Used Wrong Data ☐ Formula Error
	☐ Selection Error ☐ Other
Infrastructure	☐ Memory ☐ Space ☐ Other
Other	Default sort order for PROC REPORT (FORMATTED) is different from other procedures (INTERNAL).

Issue Log - Format Creation Technique

Date: 5/13/03	Project: IT Resource Utilization
Initiator Name	D. Blakeley

1. Description:

I needed to associate a project type with each project name in the Analysis Dataset (RUDBM02A.sas7bdat). This was required for reporting purposes. I had a Master Data file (project.dat) that contained the project type code for each project name. It would be easy to read this file into a SAS data set, then sort and merge it with the larger transaction data set. However, in looking for a different approach that would not require a sort and merge (which is useful when dealing with very large data sets) I discovered that a user-defined format could serve as a table look-up much more efficiently.

2. Turnover:

Issue Passed To: File	Date: 5/15/03

3. Resolution:

Create a permanent user defined format that associates a project type code with each project name. This requires DATA step processing that becomes input to PROC FORMAT (see RUDBM01A.sas). Then use a PUT statement to create the project type variable (see RUDBM02A.sas).

4. Return:

Issue Received From: n/a	Date: n/a

5. Classification - check all that apply:

Data	☐ Missing Values	☐ Wrong Dataset	☐ Incomplete Data
	☐ Different Format	☐ Unknown Values	☐ Other
Coding	☐ Faulty Logic	☐ Used Wrong Data	☐ Formula Error
	☐ Selection Error	☐ Other	
Infrastructure	☐ Memory	☐ Space	☐ Other
Other	Efficiency technique to avoid sort/merge with large data sets.		

Recommended Reading

Aster, Rick. 2002. *Professional SAS® Programming Shortcuts*, Paoli, PA: Breakfast Communications Corporation

Calvert, William S. and Ma, J. Meimei, *Concepts and Case Studies in Data Management*, Cary, NC: SAS Institute Inc., 1996, 156 pp.

Cody, Ron. 2001. *Longitudinal Data and SAS®: A Programmer's Guide*, Cary, NC: SAS Institute Inc..

Cody, Ronald A. and Smith, Jeffrey K., *Applied Statistics and the SAS® Programming Language*, Englewood Cliffs, NJ: Prentice-Hall Inc., 1991, 403 pp.

Diloria, Frank C. and Hardy, Kenneth A.., *Quick Start To Data Analysis with SAS®*, Belmont, CA: Wadsworth Publishing Company, 1996, 301 pp.

Kevin Forsberg, Hal Mooz, and Howard Cotterman, *Visualizing Project Management*, New York, NY: John Wiley & Sons, Inc., 2000, 354 pp.

Marge Scerbo, Craig Dickstein, and Alan Wilson, *Health Care Data and the SAS® System*, Cary, NC: SAS Institute Inc., 2001

Mason, Phil. 1996. *In the Know... SAS® Tips and Techniques From Around the Globe*, Cary, NC: SAS Institute Inc.

SAS® Language Reference: Concepts, Version 8, 2000, 554 pp.

SAS® Language Reference: Dictionary, Version 8, Volumes 1 and 2, 2000, 1256 pp.

SAS® Macro Language: Reference, Version 8, 2000, 310 pp.

Index

About the Author

Dan Bretheim is an independent consultant with over 20 years experience designing, developing, and implementing data driven decision support solutions for commercial and government organizations. He specializes in data mining, business model simulation, data analysis & reporting, design & code review, and system documentation. He is a principal consultant with db analytics, inc. and can be contacted at info@dbanalytics.com.

www.ingramcontent.com/pod-product-compliance
Lightning Source LLC
Chambersburg PA
CBHW081119170526
45165CB00008B/2486